GARRO

A LIFE IN TEN ACRES

ROBERT DRAKE

HOLLY GROVE BOOKS

First published by Holly Grove Books in March 2023
Text copyright © Robert Drake 2022
Email garronfield.rd@gmail.com
Book design by Mark Welland
Printed by Solopress
ISBN 978-1-7392367-0-0

This book is an account of a journey that started in March 2010, and that took me nowhere. It is an attempt to record how it is to *not go*: to be grounded; to set down roots; to stay put.

For Karen

To know fully even one field or one land is a lifetime's experience. In the world of poetic experience it is depth that counts, not width. A gap in a hedge, a smooth rock surfacing a narrow lane, a view of a woody meadow, the stream at the junction of four small fields - these are as much as a man can fully experience.

Patrick Kavanagh, from 'The Parish and the Universe'.

Contents

One:	From Paradise to Garronfield	9
Two:	A Wood of My Own	12
Three:	The Singing	21
Four:	White Croft Cottages	24
Five:	Holly	27
Six:	Naming Names	31
Seven:	Naming Places	35
Eight:	Caravan	39
Nine:	Outside In	46
Ten:	Neighbours	49
Eleven:	Sanctuary	52
Twelve:	The Imperative of Spring	56
Thirteen:	The Wood for the Trees	50
Fourteen:	Friends	61
Fifteen:	Hinterland	65
Sixteen:	Keeper	71
Seventeen:	Hearth	76
Eighteen:	Touch Wood	82
Nineteen:	Coppicing	88
Twenty:	Garrons	92
Twenty-one:	The Mortality of Pheasants	96
Twenty-two:	King of Birds	101

Twenty-three:	The Silence of Nature	104
Twenty-four:	The Lonely Swallow	107
Twenty-five:	Shade	110
Twenty-six:	Neglect	113
Twenty-seven:	October	117
Twenty-eight:	This Land	120
Twenty-nine:	A Brief History of Farming in Galloway	124
Thirty:	Midwinter Expeditions	128
Thirty-one:	The Food Bank	135
Thirty-two:	Living Room	138
Thirty-three:	March	142
Thirty-four:	Easter Rising	146
Thirty-five:	The Oak and the Ash	151
Thirty-six:	August	154
Thirty-seven:	Bellwether	157
Thirty-eight:	April	161
Thirty-nine:	Making Hay While...	166
Forty:	Nutting	169
Forty-one:	The Thinning of the Leaves	172
Forty-two:	A Visitation	175
Forty-three:	Of Lightness and Weight	181
Forty-four:	Roots	184

One: From Paradise to Garronfield.

At the field's furthermost corner the singing started. Here a small stream clattered between steep banks before ducking under a ramshackle watergate and on towards the sea. An equally ramshackle stone wall clambered from the watercourse and headed across the bottom of the field, smothered in places by dense brambles.

I looked back up the way we had walked, to the adjoined cottages we had come to see. They were snuggled into a banking at the topmost corner of the field, and the field itself lay in a narrow shallow valley, one of several in the vicinity running north-west to south east, carved by glaciers in the last Ice Age. To the left the hillside was partially cloaked by two woods, one of mixed broadleaves alongside where we stood, and a coniferous plantation further up. Between them a large steep field rose to where some farm outbuildings were just visible on the skyline. Over on the right a low ridge of granite, which had resisted the annihilation of ice, formed the eastern flank of the little valley.

This was Galloway, south-west Scotland, about five miles from the town of Dalbeattie and two miles inland from the Solway Firth. My partner Karen and I had driven from Paradise Cottage, our rented home near Gatehouse-of-Fleet in Kirkcudbrightshire, along the A75, through rolling grassland around Castle Douglas and down through granite town Dalbeattie. On the B-road to Caulkerbush the landscape abruptly changed. We crossed a broad moss of rush and sedge and thick stands of silver birch, then the road climbed among small hills where the bones showed through, and modest stands of conifer grew on either side. There were few houses or farms, and those that there were stood aloof from one another, a cottage at a bend in the road, or a farm tucked away down its own track. We looked out for a turning on the right with a postbox built into the wall, and followed a rough track for over a third of a mile.

At the field gateway we met the seller, Lynne Crichton. "Good, you've brought your wellies - you'll need them..." She pointed down the last bit of tractor-rutted track to the cottages, urged us to take care if we went inside, and said she had to wait at the gate for another prospective buyer. We set off by ourselves to investigate the buildings and the adjoining land.

No words had been exchanged between Karen and myself. We agreed beforehand to keep all comments and opinions to ourselves while we looked around and tried to take in as much of the place as possible. There would be time enough later for the discussion to which we had become accustomed after visiting properties for sale. So, from the place where the singing had started, we headed wordlessly back up towards the cottages, following the stream which was in places completely invisible beneath clumps of gorse, and noting how it penned a long narrow triangle of rushy ground between itself and the wall along the west boundary. The eastern side of the field was equally rushy, and presumably wet, but in between there were about four acres of indifferent pasture.

It was early March, and low sunlight could not quite cheer the dour granite facade of the cottages. Both would be classified as derelict, but the left-hand one was in best condition: it retained its front door and both window-frames, one of which had glass intact. The right-hand cottage seemed blind, with window-frames and door gone, and holes in the roof near the eaves. Inside, black holes gaped in rotted wooden floors, and ceiling plaster carpeted the rest. The twin kitchen ranges, either side of the central gable, belched soot and debris onto their hearths.

At the rear of the cottages were brick extensions, most with their flat roofs still intact, but one had caved in, and a well-grown elder tree was thriving on the rotted remains of the roof timbers and floorboards, and when seen through the side window, looked uncannily like an exotic specimen in some botanic glasshouse. The overgrown gardens belonging to the cottages were still vaguely delineated by tumbled stone walls and a few straggly hawthorns. In the corner of the left-hand garden a large spruce towered above the chimneys, its lower branches spreading for several yards in all directions, so that the tree had to be circumnavigated. At the top of the bank behind the cottages a ruinous stock fence just about marked the division between the buildings and the neighbouring field.

The estate agent's map indicated that this fence was to be

redundant: in order to provide the cottages with a more adequate curtilage to the rear, an odd-shaped section of the field had been appropriated. There were no marks on the grassy slope, but we could roughly work out where the new boundary would be. The cottages and their land were one of four lots in the sale of Auchenlosh Farm: the farm and surrounding land was the largest lot; a block of four fields lying behind the cottages was next largest; the lot we had come to see was the smallest and, to us, the most desirable.

All the lots were for sale by sealed tender, the peculiarly Scottish procedure whereby you work out how much you are willing to pay, make your offer in writing, and hope that it is enough to secure the property. Although sellers are not obliged to accept the highest offer, it would be inconceivable that this would not happen in most instances.

"Have you seen enough?" I asked, and Karen nodded. "I think so". As we made to leave I took stock of what we had seen: the two cottages, well-built originally but now in such a state they would need complete renovation; ten acres of poor land, four-tenths pasture, six-tenths rushes; a secluded location in attractive surroundings, situated at the end of a long track, almost all of which belonged to someone else.

On the journey back to our rented cottage there was silence in the car, each not wanting to be first with the inevitable question and its equally inevitable response.

"What do you think?"
"No, what do *you* think?"

Two: A Wood of My Own.

How had we come to be in that place, at the beginning of March 2010? Apart from the already-mentioned prosaic journey along the A75, the answer to the question requires a backward leap of a little over ten years.

In the final September of the old millennium Karen and I moved to Blencogo in north-west Cumbria, a pleasant and quiet village with a pub and a village hall. We soon found it to be the best place we had lived in our time together. I was self-employed, my speciality stonework, and Karen was a landscape architect for the environmental charity Groundwork. Before long we had made several good friendships in the village, and settled into the rhythm of small community life.

A few months after we arrived, the opportunity came to rent a small plot of ground near the village hall, from our neighbouring farmer Mike Ferguson. This was the chance to start growing willow for basketmaking, an idea that had been germinating ever since Karen had looked into the cultivation of willow as part of a Groundwork project. Mike also had a range of low outbuildings that he was willing to rent out, and all of a sudden I had the means to establish a sideline in willow production, which in time would further expand to include a firewood business.

My father died unexpectedly on August Bank Holiday Sunday 2000, and once my mother and I had dealt with the immediate aftermath, there came a new routine of monthly visits to Warwickshire to help out with the things she was increasingly unable to do for herself. I fetched and carried and lifted, but most of all it was a 700 mile round trip to cut the lawn. Her farmer landlord had offered to fence off the garden and return it to grazing, but there were too many memories of my father's labours growing food on that little plot. After he died we harvested the remaining vegetables, then I levelled the beds and let them grass over, but it was too soon for sheep to graze where she had so often seen him crouch to plant out seedlings, or carefully hoe along his neat rows.

So for three years this monthly journey became a ritual. I would leave early on Saturday morning, and pick up fish and chips for our lunch, in Wellesbourne, a few miles from mother's. I'd leave after the Sunday roast she insisted on making, even though it was an increasing effort to do so.

Then our friend Annette, who lived in the cottage adjoining ours, told us she was intending to sell. Although I had been 'in business', self-employed for most of my working life, it is Karen who has always been the more financially astute of the two of us, and without hesitation she let Annette know we'd be interested in buying her place. As well as an investment, she saw this as the chance to solve the long-distance commute to Warwickshire, if only my mother would be prepared to leave her long-time home and all her friends.

I broached the matter on my next visit. She didn't even hesitate, agreeing with some relief, as it turned out she had been very concerned about the amount of time I was having to spend in looking after her. We made a private deal with Annette, and the sale went through quickly. By the end of May 2004 my mother was settling in next door. My life for the time being was rather easier than it had been.

By mid-decade my life was not only easier but truly comfortable. I had plenty of work - stone-walling, supplying firewood, and growing willow - and life in the village was more than pleasant. I had a loving partner, a comfortable house, a fine garden where we would dine out on summer evenings and look across the Solway to the Galloway hills. We had good friends, regular village events to be part of, and even more regular Sunday lunches at the New Inn. There seemed to be no unfulfilled ambitions gnawing at me: the kind but firm anchor of my mother's presence next door meant that we were settled, and that didn't seem to trouble me at all. In the run-up to my 50th birthday I did a kind of stock-take of my circumstances, evaluating the past and considering the future. Whatever the future might have in store, the word that most accurately described my present circumstances was *contentment*.

It didn't last long: less than eighteen months. My mother died unexpectedly on the first day of 2007 - her heart, like my father's, had given up - and I was hurled once again into the strange dislocated world of dealing with important practical matters whilst knocked sideways by grief.

Karen and I tried to get back to normality, but things were not the same as before. With mother's death there was no longer that comfortable anchor to keep us put, but we stayed. We refurbished her cottage and installed a tenant. I was adrift, but could see no welcoming shore on which to land.

It is as difficult to chart the development of an idea as it is to see the growing of a tree. Even vigorous hybrid willow, reaching twelve feet in one season, does not grow perceptibly before your eyes. But go away and come back in a few days, and the buds on each stool will have burst and become two inches of delicate stem and fresh leaves. Then the stools will vanish beneath a cluster of stems, all trying to outcompete the others and grow tallest. Then waist-high, and shoulder-high, the crop sways elegantly in the breeze.

I can't point to any single moment of revelation. All I can say is that at some point during the late summer of 2008 a fully-formed idea was present in my head, that could be expressed in clear and simple words - *I want a wood of my own*. And I knew exactly what sort of wood it should be - a coppice, harvested on a seven-year rotation, providing fuel for heating and cooking.

Two events that I can recall played their part in the development of the idea. Earlier that year I had done some walling work on a small farm overlooking the Vale of Lorton, south of Cockermouth. There was some established woodland, but of greater interest were the woods that the landowner had planted himself, and was enjoying as they grew towards maturity. I felt a keen envy of him in his delectable spot.

The second event occurred while walling at a farm north of Cockermouth. The farmer was selling up, and wanted his walls tidying up prior to the auction date. I had worked previously for this man, and at one stage expressed an interest in a wood on his farm, an overgrown coppice of several acres. He did not take up my offer of renewing the coppice, but he did tell me, almost with a sense of pride, that in all the twenty or so years he'd owned the farm, he'd never once set foot in that wood.

On the tidying-up day I finished-off along the back of the wood, and discovered a resurgence of the original perplexity I had felt at his indifference to the place. As the afternoon wore on perplexity honed itself into pure rage as I thought what I could do with such a wonderful resource.

One important stage early in the idea's development was a recognition that Karen was not entirely content with living in Blencogo. Although it was the best community we'd ever known, something about the surroundings was less than satisfactory for her. It was too flat, too agricultural, without enough accessible wild places nearby. Her heart desired a more varied landscape to live in. And now I wanted something that the place could not provide.

On Saturday 4th October 2008 Karen and I sat down at the kitchen table for lunch. Once the tea had been poured, without any preamble I said: "I've got something to tell you." I saw her eyes widen slightly and her face stiffen. I could imagine all the 'bad news' thoughts that would be racing through her mind, so I blurted it out: "I want a wood of my own..." Her relief was palpable: she was puzzled and intrigued. "Okay", she said cautiously after a moment or two. "What's brought this on?"

I told her about my recent experiences on the two farms, and how they had helped to crystallise the flake of a thought that had been growing inside me for some time. I reminded her that she'd said on more than one occasion that Blencogo was not perfect. I suggested that since it could not fully meet our desires, we needed to look elsewhere. "It's time for us to move on", I said.

This was the opening scene in a long drama of excitement and upheaval. Before we moved to Blencogo we had been looking for somewhere with 'a bit of land', perhaps a couple of acres, enough to keep a horse - another of Karen's heart's desires - and somewhere for our energetic border collie to run about. We'd looked at a number of places in north Cumbria, and just over the border, but nothing had been quite right. The nearest was perhaps a cottage near Hethersgill, with a barn and two acres of beautiful hay-meadow. But cottage and land were on either side of a minor road, and the cottage was ugly to our eyes. Although the house in Blencogo did not have more than a decent-sized garden, we decided to take a look, fell in love with it, and that was that.

Now I knew that I wanted a wood, and Karen's horse-dream was still very much intact. But a wood would need more space, perhaps five or six acres. Through research I found that to be self-sustainable in fuel a typical house would require seven to eight acres of coppice woodland, so our sights became set still higher.

Ten acres. We spoke the words with a kind of amazement, never having thought that we might aspire to owning so much land. We began to think about where. The Lake District was beyond our reach in money terms, and in fact in recent years I had begun to find all too many parts over-busy. We looked north, beyond Carlisle, to the big tract of country in Cumbria and into Scotland, that we had explored before Blencogo. Being mostly upland, some parts of it appealed to Karen. I favoured staying in Cumbria, but her sights were set across the border. In the end a course was set out for us, when in midsummer 2009 Karen applied for and accepted a job based in Kirkcudbright, 35 miles away due west, but 80 miles away by road.

This was a lot for a daily commute, and straight away she began to look for somewhere to rent. A colleague said she'd seen a place advertised in the local Co-op, and this turned out to be Paradise Cottage, on Glen Farm in Skyreburn, west of Gatehouse-of-Fleet. It didn't take a lot of thinking about: the cottage was in an almost-idyllic location, and Robert the farm manager hinted, in response to Karen's enquiry, that there might be grazing available for ponies. That clinched matters, and we agreed to take on the tenancy, although the cottage was not available until mid-November.

When we made the decision to move on, we left matters until the new year while we thought things through. But once 2009 got underway, events moved at some speed. We put both houses on the market, and mother's cottage - as I thought of it - was sold by early April. Once Karen had taken on her new job, and had found us Paradise Cottage, we could be more committal towards potential buyers of our house, and soon we accepted an offer, although in nerve-wracking fashion nothing was settled until a week before we were due to move.

In mid-September we visited Jane Barker's Fell pony herd in Heltondale, south of Penrith. We had been talking about horses and ponies ever since decision day, and were inclining towards the all-rounder breeds of sturdy pony, such as Dales and Fell. Jane took us on the back of her quad bike to where the main herd were grazing their way through an open-plan of fields. We'd come to the conclusion we'd want to have a pair, as company for each other: a four-year-old that Karen could work with right away, and a foal to follow on. In the fashion of putting the cart before the horse, we were on our way to putting the pony before the pasture.

"You don't choose them - they choose you", Jane was to tell me later. And so it was that Finn, a four-year-old gelding, chose to spend a good deal of time with us, while one by one the others lost interest and drifted away to resume grazing. He seemed gentle and steady, with kind-looking brown eyes. Then we headed downhill to the field with the foals and the stallion. Here decision-making was more difficult, since all the foals were diffident and wary, a bit white-eyed when we got too close, and there wasn't much, to our novice eyes, to distinguish them. Eventually I pointed out one, named Jiminy, that I liked the look of. But I said to Karen that I'd only go for him if he'd let me approach him and handle him.

At first it seemed this wasn't going to happen, but eventually he did let me approach, and there are two photos Karen took of me reaching out to touch his head and scratch behind his ears. That evening I phoned Jane and said we'd like to buy Finn and Jiminy, provided she would be prepared to keep them until we'd sorted out pasture for them. She asked how we'd decided on Finn from his peer group, and I pointed out that apart from anything else he'd spent the most time with us. "You don't choose them - they choose you", she said.

November 14th was our moving day, when the last strands of contentment with life in Blencogo would finally give way. A month earlier had been the first of a series of 'end of an era' days, when I finished my last day building dry stone walls in Ennerdale, where I had worked for the National Trust and individual farmers for almost 15 years. I drove up to Bowness Knott, where the metalled road ends and the forestry begins. When I was rebuilding a wall in Silver Cove, high up on the southern side of the valley, I had permission to drive on along the forestry tracks to the point where I had to park up and begin a three-quarter hour's climb to the wall. This evening I could only turn around in the carpark and head back down the valley for the last time.

I went home via a detour over Fangs Brow into Loweswater, then through the Vale of Lorton. In all these places I could see extensive stretches of wall that I had rebuilt over the years, and the sight gave me a deep sense of satisfaction tinged with melancholy. Three weeks after that it was the turn of Wythop, between Cockermouth and Bassenthwaite, to hear my last farewells - another place where many tumbled-down stone walls now stood proud once again.

Two days later I went to the little graveyard at Bromfield Church, just down the road from Blencogo, in time for the silence at 11am on Armistice Day. It was a calm and sunny morning, and I stood by my mother's gravestone, with a full heart, as a skein of pinkfeet geese flew high overhead towards the Solway Firth. The other side of that gulf of mud and salt water was where I was headed in three days' time. On my way out of the graveyard I paused one last time to read - though I knew it by heart - the inscription on a bench by the wall: 'Those we love are with us still, because love itself lives on.'

If Karen had at first been anxious about the idea of moving on, now it was my turn to feel the terror of the unknown. She had a job to ease her into place: I had nothing of the kind. My son Peter had taken over my walling contracts in Ennerdale, Wythop, and at Watermillock near the foot of Ullswater. Once we had settled in to Paradise Cottage I was able to get some work with him from time to time, though they were long days, setting off at 6.30 am, and getting back thirteen hours later.

Nevertheless that helped to tide me over, and before long some work in Galloway began to come in. I advertised, and the best advert - word of mouth - started to have an effect. I also had a crop of willow still waiting to be harvested in Blencogo, and Karen would come with me on those days, so that we could get a decent amount cut and piled high on the back of the pickup.

Finn, and the foal formerly known as Jiminy, but now called Mister, arrived at Paradise at the end of the following January, the fulfilment of

Karen's 30-year dream. They should have been with us before Christmas, but winter froze the high ground, and Jane couldn't get out of Heltondale safely with a trailer. We had to wait for the thaw, and the freeze lasted a full month. Their grazing was a large steep field behind Paradise, and at the top was a disused sheepfold with seven-foot high walls, which I'd partially roofed and enclosed as a loose-box for Mister, who would need night-time shelter and extra feed through his first winter. I had extended the loose-box roof to provide shelter for Finn if the weather was bad, but he scorned this luxury, instead spending every night outside the sheepfold, irrespective of the weather conditions, standing as close as he could to his young half-brother, as could be seen by the slur of hoof-prints in that one place.

From very soon after we had arrived at Paradise we were scanning the local paper and estate agents' windows for properties for sale with the extent of land that we wanted. None that we looked at were suitable. One consisted of a habitable though uninspiring house, a building plot next door, and a couple of small fields. There wasn't enough land, and anyway the fields stretched down to the A75, the only trunk road from Europe to Northern Ireland. One of the unsatisfactory things about Blencogo, especially for Karen, was the amount of traffic noise - not along the street, but along the A-roads that bracketed the village. This was especially noticeable on still evenings, and tempered the pleasure of sitting-out when the weather was good.

Another property we looked at had more land than we needed, but it was high up, beyond Dalry, quite remote and bleak. The outbuildings were good, but the house, which was used as a hunting lodge, was too big for us, and there was something fishy about the whole thing. The man who owned it seemed rather too evasive about some details, so we gave that one a miss, too.

Not long after that visit, one evening at the end of February, I looked through the adverts towards the back of the Galloway News, and there was an estate agents' photograph of a pair of cottages in need of restoration, together with ten acres of land, somewhere near Dalbeattie. In retrospect it seems strange that the agents should choose to highlight this smallest part of the estate that was up for sale, but the photograph stared out at me, the asking price was affordable, and a few days later we rolled along the access track and walked in our wellies down the rushy field to the place where the singing started.

Three: The Singing.

What was the singing? From the outset I have used the phrase to remind myself, and to tell others, that I experienced a moment of intense emotion at the bottom corner of the field. But not for some time after did I begin to wonder what I'd meant.

The universal language of music has always had the power to move me deeply: deepest of all is my emotional connection with what I will call the 'downturn'. This occurs in many types of music, but perhaps especially classical, and I have learnt from the sleeve-notes to Mahler's *Song of the Earth* that the 'falling second' is a device to express yearning.

In the second movement of Chopin's Piano Concerto No 2, the piano, when it enters, flutters upwards quickly like a skylark ascending on the ladder of its own song, then pauses, and downturns, and this is always the moment that strikes most surely and deeply into my heart.

Grieg's Piano Concerto was always a favourite of my mother's, and I was accustomed to hearing it throughout my growing-up. When I was sorting out music to play at her funeral, the slow movement was a strong contender, and I listened to it again one evening to make sure. The 1959 recording I have has Clifford Curzon at the piano with the London Philharmonic Orchestra, and I think my mother's version was the same. I've heard many renditions of this concerto, and in some the piano comes crashing into the *adagio* heavy-handed, which to me is utterly wrong. Curzon's first note is as pure and sweet as the first drop of rain in a still lake. Then follows a sequence of phrases, each downturning at the end. That evening, listening more intently than ever, by the time the piano repeated that first pure note, my heart was full of grief and my eyes full of tears.

Beyond the classical realm, Beth Neilsen Chapman's song *Every December Sky* is inextricably linked, for me, with the tragic death of grandchildren, and it is the music I turn to whenever there is occasion to recall such loss. Throughout, it is the scalpel of the downturn that lays my emotions bare.

There are so many other instances that have the power to move me to the core: Bill Evans' *Peace Piece*; the opening piano chords of Mary Chapin Carpenter's album *Time Sex Love*; June Tabor's *Maybe Then I'll be a Rose*. But to complete something of a circle - until I listened more closely, I'd thought that in Vaughan Williams' *Lark Ascending* the violin kept going higher and higher until beyond hearing. But in fact it does rise as far as it can into one final ethereal note, then downturns. The lark is not, after all, an immortal bird, but must in the end descend to earth, to his mate and her nest.

I've received a few strange looks when telling others about the singing, in fact my son Pete was bold enough to inform me I was mad, and I think he was only half-joking. But I know I'm not unique, that others have experienced something similar, although expressed in different ways.

In Peter Scott's *The Eye of the Wind* he describes the spine-tingling moment of identifying a lesser white-fronted goose for the first time in Britain, the same feeling he got listening to the slow movement of Sibelius's Violin Concerto. Tim Smit, founder of the Eden Project, writes about his own moment of recognition, in his book *Eden*. The instant he saw Bodelva Pit he was sure it was the place, and felt as though he had been there before. He knew he had found his Eden.

But closest to my own experience is that of Katherine Stewart, as recounted in her 1960 memoir *A Croft in the Hills*: She and her husband spotted an advertisement for a seven-roomed house, with forty acres of arable land and an outrun on the moor. Stifled by city life, it was just the sort of place they had been looking for. They got out the map and found Abriachan, among the hills above Loch Ness. 'Music was sounding in our ears.'

Katherine, together with husband Jim and young daughter Helen, go to visit. Later, returning to Inverness, they felt that the house on the hill was speaking to them, and they listened in silence to what it had to say.

Now that I have set all this down in some kind of order, two things become apparent. The first is the repeated reference to my heart, to the core of my being, and yet it was my head, not my heart, that heard the singing. The second thing is that, for the most part, the examples of emotional connection refer to instrumental passages rather than singing. The curious thing is that neither solo nor choral singing featured in my experience. I did not hear voices.

So in the end, after all the conjecture about the way music touches, the singing was actually the fizz of raised blood pressure as my heartbeat quickened. For there I was, trying to be objective and rational and taking careful note of all that I was seeing (as we had agreed, Karen and I, beforehand), while my pulse was racing faster because I knew, although it would be some time before I could say it, that we had found what we were looking for. It was my blood singing.

Four: White Croft Cottages.

We made an offer that was well above the asking price. We wanted to give ourselves the best chance of owning the place. By the time we reached that point we had spent hours going over the pros and cons, which, I recognised, Karen needed to do far more than I did, because she is not impetuous. On the rare occasions when I go shopping for clothes I will usually buy the first things I see that I like, even if they are all in the first shop. Then I go home. I never accompany Karen when she is buying clothes.

On a sunny Friday afternoon at the beginning of April we returned to Paradise Cottage to find the phone ringing. It was our solicitor, to give us the news that our offer for White Croft Cottages had been accepted. He suggested that we might celebrate with some bubbles, but perhaps keep the proper champagne for later...

We did indeed celebrate, with malt rather than bubbles, then settled down to three months of ironing out snags, sorting out services, and making plans. On Friday 25th June we took possession. I cut the wire that had kept cattle away from the buildings and declared White Croft Cottages 'open'. We planted a young Laxton's Superb in the dream of an orchard, and poured a libation of Ardmore malt on both doorsteps and around the apple tree. Then we drank a toast to the future, walked the boundaries, saw ringlet butterflies and purple orchids. At the Singing Place, Sky, our border collie, found the stream, as he had done on that first day, and joyously splashed about in it.

The first job was to peel back the tangled mat of neglect that had grown up around the cottages. Long grass mowed, scrub cut down, the big conifer limbed so that we could walk underneath. Gradually the cottages emerged from their wilderness, acquired their own curtilage, and a bolt once more on the surviving front door. We had to make provision for bats, who had been roosting in the roof space, by putting up a bat box for the time when the roof was taken off. We had to exclude badgers, who had set up a 'day bed' under the floorboards in one back bedroom. A local contractor cut our field for hay. We'd hoped for small-bale hay, but the weather broke too soon, and the crop was baled and wrapped - 32 big black bales stacked at the bottom of the field. And we welcomed a succession of visitors, both family and friends, cooking (on the barbecue) what was most likely the first food for forty years - which was when the cottages had last been inhabited.

In the background there was much activity of the administrative sort, applying for planning permission and a building warrant. No work could be done on the buildings until all this paperwork was completed. In the meantime, on our visits, we paced around the place, imagining where things might go.

The land divided itself quite naturally into the right proportions for what we wanted to do. There were four acres of thistly pasture, and a little less than six acres of rushy bog, which would be perfect for planting trees. In the south-eastern corner of the field there was a low-lying spot, dense with rushes, which would be ideal for a pond. Between our main gate and the cottages was an area that could accommodate a range of outbuildings, and the first and smallest of these, a 12ft by 20ft timber workshop, was ordered from a sawmill near Newton Stewart.

In mid-October I took delivery of what seemed to me a massive digger, a 13-ton JCB with three-foot wide tracks. This machine was to prepare the rushy ground for tree-planting next spring, and the wide tracks meant that, despite the digger's weight, it would not sink into soft ground. Well, not too much anyway. I had never done any 'dolloping' as it is called, and the technique took some getting used to. The idea was to dig out a bucket-sized turf of vegetation, and turn it upside-down next to where it had been dug, forming a low mound into which a tree could be planted, with a shallow pit alongside to help with drainage.

Two months before I'd hired a 3-ton mini-digger to dig percolation test holes for our future drainage system, and to carve out a base for the workshop. The last time I had used any kind of excavator was forty years previously, and I was a little more than rusty. Operating a digger is rather like trying to pat your head with one hand and rub your stomach with the other. It's as well I was working alone, as no-one else was endangered by the sudden dips and lurches as I wrestled to master the controls.

Sitting high up in the big JCB it was sobering to think what damage

I could inflict if I wasn't careful, and I resolved to be very, very careful. Nevertheless my first couple of hours' work had mixed results. Mostly my dollops were small hills rather than low mounds, as I tended to dig too deep. Every now and then I'd get one right, and think I was getting the hang of it, but then the next one would be too big, or the bucket-full would come out right-side up, and I'd have to fiddle about to turn it over.

It took four days to dollop six acres. By the end of the last day I was getting better at it, but then this was a job that I was never going to do again. I did manage to keep the machine afloat, although there was one place, not that far from the cottages, where my heart beat faster as the tracks sunk into the boggy ground, almost disappearing completely. But the digger kept going, and I was saved the ignominy, and expense, of having to call for a bigger machine to get that one out of the swamp.

At the beginning of December I paid a final visit for that year. A sense of melancholy prevailed, as loss had come to us, tragic and unexpected. The place was blanketed with snow: the only greenness was around the spring nearest the cottages. The dog disturbed a pair of snipe from out of the green circle, and their jinking flight took them fast away down the field and out of sight.

Five: Holly.

The Mosedale Holly Tree, near Loweswater in the western Lake District, is one of a rare kind of tree: it is named as an individual on the Ordnance Survey 1:25000 map. I don't know how many other trees in Britain have this distinction, but of the maps I am most familiar with, covering north-west England and south-west Scotland, I am fairly sure the Mosedale Holly is unique.

It grows in magnificent isolation on the western flank of Mellbreak, guardian of its little valley. I had seen the tree a number of times from a distance, and its name always seemed to stand out whenever my eye roved over that portion of the map. Years of such aloof familiarity passed, until something of an imperative entered my mind, and I knew I had to make a visit to see this rare tree at close hand.

It was midwinter, and the day was not ideal, with low cloud and a rawness in the air. As we climbed higher in the valley there was frost on the rushy ground. We almost missed the tree altogether, as it stands some way below the track to Scales Beck, and the mist did a good job of concealing it. But a ghostly patch in the greyness gave it away, and after a short steep scramble down, the Mosedale Holly Tree came into focus.

This ghost-tree became ever more wraith-like as we approached. Its evergreen leaves were frosted, as though it was camouflaging itself within the mist. But on closer inspection we saw that the spikes on every leaf bore a starburst of ice-splinters, turning a normally prickly tree into an ethereal bastion, glittering white blades turned outwards in every direction.

"When I have a wood of my own," I said to Karen, "I'm going to plant a holly grove in it." She smiled mockingly at me. "Just how big is this wood of yours going to be?"

This happened in the limbo time between us knowing that we needed to move on to find somewhere with that 'bit of land', and us actually realising how much ground we needed to fulfil our dreams. But in the end a holly grove would prove to be the starting-point for what would be that 'wood of my own'.

From the earliest days of my working life I had loved being among trees. In the early 1980s I spent the best part of a year as a sub-contractor for the Forestry Commission, felling and extracting timber from a steep section of Norway and Sitka spruce at the very northern end of Grizedale Forest. At a time when many forestry contractors were going out of business due to the low price of pulp timber, I and my business partner were trying to make it work by a low-input strategy: three second-hand chainsaws, and a borrowed horse called Ginger, a fifteen-year-old Suffolk cross Clydesdale who had worked in the forestry all his life, and knew far more about it than we did.

Once we stopped fighting with him, and learnt to allow him to do what he did best, the work became much easier. It was still hard, dangerous, and financially unrewarding, but it was one of the best jobs I have ever had. When eventually debt crushed our little enterprise, I still longed to work among the trees, and sought every opportunity to do so.

Spool forward more than twenty years, and I am living in Blencogo, with a business sideline in providing firewood. Supplies proved erratic at first, until I had permission to take firewood from Parsonby Wood at Plumbland near Aspatria. It was in the middle of this wood that I came across a holly grove, a natural circle of low trees among the understory, possibly formed by a single tree with downcurved branches that touched the ground and took root. I was careful not to damage the circle when felling other trees nearby. It seemed to me a remarkable feature: I kept the image of it with me long after I was denied access to Parsonby Wood, and it was rekindled in bright colours as we gazed at the Mosedale Holly Tree.

During the winter after we bought the cottages and land, we planned what sort of woodland we wanted to establish. Most of what we would plant would be coppiced, with just a few trees per acre left to grow on to maturity. Our aim was to provide fuel for heating and hot water and outdoor cooking. In the end we decided on a traditional oak/ash/hazel core, with alder in the wettest places, and a sprinkling of downy birch, rowan, and hawthorn. Karen was keen that the visible woodland edges should be attractive and fruitful, and so we added a list of what I mockingly described as her 'exotics' - guelder rose, field maple, crab apple, aspen, damson, dogwood, among others. There would be flowers in spring, berries and bright colours in autumn. My chainsaw would not be allowed near these trees.

We planned two areas of planting, and tentatively gave them names. The long triangle of rushy ground along the western boundary we called 'March Wood', because the march dyke, or boundary wall, that separated us from the neighbouring farm, lay alongside it. Almost all of this area was dolloped, except for a narrow strip along the stream. On the other side, below the cottages, a number of springs bubbled to the surface, and so we called this area 'Spring Wood'. There would be nothing obscure about our naming of things.

As winter drew to an end, preparations were made. The 12ft by 20ft timber workshop was made ready to store 3300 trees and assorted stakes and shelters. A family workforce was booked for a weekend visit. Tools and equipment were cleaned, sharpened, oiled. And the first day of planting in March Wood was centred around the making of a holly grove.

It was a bright cold day in late February 2011. There were adults in strange hats with flaps and tassels, and their children, dressed predominantly in pink. My family worked together that day to plant and stake and fit shelters on many trees, but at the heart of their work, and at the centre of our thoughts, was the holly grove. I had planned it to be about sixteen feet across, giving a circumference that would allow thirteen trees to be planted - one for each year of a young life - and a space for an entrance facing just a little east of south. At the centre of the circle we planted a single oak.

In the following days March Wood grew in extent, with eventually more than 1800 trees planted, interspersed by willow cuttings. The holly grove gained a gently curving pathway to its entrance, an avenue of thirteen rowans. This would be a place to visit frequently, to see the young trees lifting their waxy leaves towards the light, and imagine in a future time how selected branches would be trained across the circle to form a canopy.

If a yearning thought could take flight from out of the holly grove, in a straight line on a compass bearing of 162 degrees, just a little east of south, it would soar over Hawthorn Hill, then down to the shore where Southwick Water spills onto Mersehead Sands. Angling across the shining band of the Solway, the thought would make landfall on the Cumbrian shore between Northside and Siddick, then follow the coastline, a few miles inland, with Black Combe in front, and over that mountain and Duddon Sands beyond, to a cemetery at Hawcoat, where a lovingly hand-crafted headstone bears the inscription:

> Searlzy
> JORDAN JOHN
> WALDERS SEARLE
> 24.02.1997 – 05.11.2010
> forever in our hearts

Six: Naming Names.

The place was not called Garronfield at the beginning. The buildings were known as White Croft Cottages, with an innominate parcel of land, in fact a single field. It is quite possible that the field once had a name, but that knowledge has been lost. Forty years had passed since the cottages were last inhabited.

From the start our intention was to turn the cottages into one, so that very soon 'Cottages' would become singular. The field would become a smallholding, sub-divided for pasture, woodland, willow beds. At some stage, quite early on, I had the inkling that the land should have its own name, as it was a holding in its own right, rather than being part of something bigger.

White Croft Cottages had been part of Auchenlosh Farm Estate, which in turn had once been part of Auchenskeoch Estate. The fragmentation of small to medium-sized estates seems to have been a pattern in southern Scotland, with just the largest being immune to economic ups and downs. But we were the beneficiaries of such a fragmentation, and now owned the smallest of the four lots of Auchenlosh, not land enough to be viable as a farm, but just right for what we wanted to do.

I wanted to find out as much of the history of the place as possible. The cottages were not marked on the 1854 Ordnance Survey map, but were on the 1895 edition. I trawled the local history shelves in Dalbeattie library, gleaning what I could from any number of sources. One afternoon a new book appeared: *History of the Colvend Coast, Napoleonic & Victorian Times*, by John Gillespie. I pulled it off the shelf and opened it at random near the middle. A photograph of White Croft Cottages stared out at me.

When I got the book home and began to investigate, it transpired that the Gillespie family, who had farmed for generations at Drumburn, Colvend, had a direct connection with the cottages. Gillespie sisters Sarah and Agnes married Buchanan brothers Alexander and William. In 1870 these couples were the first tenants of newly-built White Croft Cottages on Auchenskeoch Estate. The cottages were probably constructed during the building and expansion of Auchenskeoch Lodge, and the granite masonry, particularly along the front elevation, is of a far higher quality than might be expected for farmworkers' dwellings.

Some time after finding out about the first occupants we had a visit from Cameron Patterson of Dalbeattie, whose father (also Cameron) was a self-employed farm worker, and the family lived in both of the cottages (firstly in the right-hand one, and later in the other) until Cameron junior was nine years old. They left in 1969, and were the last people to occupy White Croft Cottages. Cameron said that he remembered it to be a happy place to live as a child.

So we knew who the first and the last occupants had been, and kept our ears open for further clues as to who lived in the cottages in the intervening years. In the meantime we uncovered more names and dates, not of tenants, but of those involved in the building and maintenance of the cottages.

I was busily demolishing a brick partition just inside the front door of No 1 (we had decided to temporarily call the left-hand, northern cottage, No 1, and its southern neighbour No 2, just to save confusion), when something unusual caught my attention. The bricks were red clay, with two horizontal holes running through them, rather like double-barrelled drainage tiles. Some of them were 'brick-and-a-half' in length, and I noticed that one of these, having fallen to the ground, had a piece of wood protruding from one end.

I pulled the stick free, and saw that it was just a joinery offcut, a wedge-shaped piece that had been cut from a larger timber to make it fit. But then the hairs on my neck tingled as I realised there was writing on the stick. On one flat face, in pencil, was written: 'Robert McGuin, Mason, Millburn Bank, April 20th 1870'. On the narrower side there was more writing, beginning with 'go to thunder', then possibly 'yellow', and the rest indecipherable. Quite likely a comment on Robert's working relationship with his boss.

That evening I searched the map, and found that Millburn Bank was a place little more than a mile to the east. Gillespie's book then revealed that this had been the location of the Robson family in the latter half of the nineteenth century, a family of stonemasons descended from Joseph Robson, 'Master Builder'. The Robsons carried out work for both

Southwick and Auchenskeoch estates. It seemed that there were very few dwellings, especially on Auchenskeoch Estate, that Joseph did not have a hand in either building or improving. It is quite probable that Joseph Robson built White Croft Cottages, and that Robert McGuin was his disaffected employee.

A few weeks later it was Karen's turn to be demolishing internal walls, and as she broke down one of the kitchen walls in No 2, another date stick came to light. This one read: 'John Turner built this partition 1870'. Then, turning the stick over, a different hand had written: 'John Farries Rebuilt it in December 1938. Signed Charles Green, Dalbeattie' (this written over some earlier words, rendering them illegible).

These finds, together with a joiner's signature underneath a window board ('Wm Hunter, Sept 1956'), allowed us to piece together the development of the buildings: built 1870; renovated 1938, with small brick scullery extensions at the back doors; further expansion in 1956, when larger brick extensions were built at the rear corners, creating an extra bedroom for each cottage; abandoned 1969, and allowed to become derelict.

Some time after these discoveries, we had one further unexpected visitor. A car pulled up in the yard and a man in his mid-sixties climbed out. "Hello, can I help you?" I asked. "I lived in that cottage," he said, pointing at No 2, "until I was four years old."

His name was Andrew Montgomery (his wife Angela was with him), and he commenced telling me an intriguing story. His father, Anton O'Kora, was Yugoslavian, and during the Second World War the Yugoslav army disintegrated very quickly. Anton was displaced, along with many thousands of others, and eventually found his way to England, and then to southern Scotland. He was a ploughman, as Alexander Buchanan had been, and found work on Auchenskeoch Estate, where the owner, Captain Philip Barton, was very kind to him, helping him to become naturalised in 1949.

As we walked about Andrew remembered many things: a swing his father made for him, out the back; him and two older brothers in one bedroom, his parents in the other; a bull in the field behind the cottages, with a ring in his nose and a trip chain attached to it. He said the cottage was very basic, with the traditional range in the kitchen, ovens and plate-warming either side of the fire, and a kettle hooked on a swinging arm. He wasn't born at the cottage, but lived there from 1951 to 1954. His family moved on just a couple of years before the final improvements might have made the cottages a little (but not much) less basic.

Being farmworkers' cottages, it is more than likely that many tenancies were of short duration. It is easy to speculate that, over the course of a century's habitation, a hundred or so people might have lived in the two cottages, however briefly. And there will have been births, and deaths, and all the rest of life's pageant.

Now we were about to do something new. For the first time in 150 years White Croft Cottages would become a single habitation, and a permanent one, a place where Karen and I could live out the rest of our lives. Scottish people often refer to the place where they 'stay'. When I first heard this I thought the person meant a bed-and-breakfast or hotel, but it soon became clear they were referring to their home. I rather took to this terminology, it had a resonance with my own desires. I could *live* anywhere, but to *stay* is to settle in one place. And that is exactly what I wanted to do: to settle, to stay put.

Seven: Naming Places.

I would as soon read a map as read a book. Maps have always drawn me, and I have loved drawing them whenever the occasion arose. Around the age of fifteen I thought my dream career would be to work for the Ordnance Survey, and I embarked on a trio of A-Levels that ought to have been the perfect combination for such a future. Sadly, one of the three subjects was Maths, and after the first term I went to the kindest of my teachers and told her that I hadn't understood much of what had been taught. "I know," she said. "Go and do Art - that's what you really want." And she was right.

Having my hopes of becoming a cartographer dashed did not diminish my enthusiasm for all manner of maps and charts. I brought that enthusiasm to our new place, and from the outset would think of it in terms of a map. We had the plan that was part of the Land Certificate registering our legal title, but this was fairly basic: outlines of the buildings and their gardens, the boundaries, the stream, and a wavering dotted line that seemed to mark the division between pasture and wet ground. The map I had in my head would have much more detail on it.

Already we had marked out March Wood and Spring Wood, and it turned out that our stream was actually the headwaters of Caulkerbush Burn, named as such on the OS map. So the English word 'stream' was banished, and the watercourse became henceforth 'the burn'. There were other watercourses of a lesser nature: a ditch that followed the march dyke which was the boundary between ourselves and New Farm, and a gutter down the middle of Spring Wood that I had dug when I was preparing the wet areas for tree planting. The term 'sike' is perfect for such small streams flowing through marshy ground, but it was not a local word. On studying the maps I found that 'sike', common in northern England, did spill over into eastern Dumfriesshire, and so I decided to apply the term, as though we had picked it up on our journey from Cumbria and brought it with us. Far Sike and Spring Wood Sike appeared on my mind-map.

The first structure I built was a solid crossing of the burn, actually a short culvert consisting of two lengths of large diameter drainage pipe, laid on the stream bed and then covered by stone and hardcore. The result was a concave bridge that dipped down from one bank and climbed to the other. I remembered that there was a similar construction, albeit of concrete, across the river at the head of Ennerdale Water in Cumbria. That was known locally as 'Irish Bridge', for reasons unknown, and I appropriated the name for our crossing-point.

The field had been ploughed regularly in times past. The evidence for this was the piles of granite boulders that had hindered the plough and had been laboriously dragged out of the way. There was a considerable mound of stones covering the tongue of ground at the confluence of Far Sike and the burn. On the other side of the field, about halfway down, was a group of large boulders in a hollow at the margin of the marshy ground. When Spring Wood was enclosed, the boulders sat in an angle of the fence, and became Corner Stones.

Many other names appeared gradually on the map as we transformed the land. Where the outfall from the septic tank cut through Spring Wood we left a strip unplanted, and this became Pipetrack Avenue. Each year I established a new willow bed, and these were: Top Shelter Belt, along the north-western boundary; Middle Shelter Belt, dividing the field in two; The Triangle; The Pollards. Karen planted fruit trees across the slope behind the cottage, and this naturally became Orchard Bank. She also wanted to enhance the richness of plant-life in what was now the bottom field, and when we dug a trench for drainage across the pasture, we deliberately carted away most of the topsoil so that the resulting scar would be less fertile, ideal for sowing wildflower seeds. Karen ordered a bespoke mix of species from a seed company, and when the package arrived (we were still living at Paradise Cottage) we found that they had nicknamed the seed collection 'Paradise Meadow Mix'. So the bottom field was gifted its name, and the top field, after much deliberation, became, well, Top Field.

But there still remained the matter of a name for the whole of the place. I began to study the local OS maps, together with glossaries of place names. It became apparent that farm names beginning *Auchen-* (as with Auchenskeoch and Auchenlosh) were common in Galloway, and especially along the Colvend coast. It transpired that this is an anglicization of the

Gaelic *achadh na*, 'field of', and although I didn't want to use *auchen*, 'field' would be a fitting nod to the historic names, while being locally distinctive, as there are just a handful of placenames in the area with that word in them.

So, having gained approval for that idea, the search was on for a term to partner 'field'. We went back to the glossaries, and in one published by the Ordnance Survey, Karen marked a number of terms in Gaelic that seemed to have resonance: *fuaran*, well or spring; *larach*, site of ruined house; *seileach*, willow; *caiplich*, place of horses; *each*, horse; *marc*, horse...

The pattern was obvious and inescapable. From the beginning of our interest in Fells, we had known that ponies of this sturdy type were called 'garrons' in Scotland and Ireland, from the Gaelic *gearran*. And so the last piece fell into place: Garronfield it would be.

I felt like a pioneer, with both the thrill and responsibility of naming things that had previously been innominate. It sent me back again to the OS Explorer map of our area, and noted once more the named woods in the valley - Whitecroft Wood, Bennel Wood, Clonyard Wood, Jockleg Wood. I wondered how long it would be before two extra patches of green appeared on the map, but probably they would be too small to bear their own names. Then something caught my eye: a mile and a half to the west, on the hill slope below Drumstinchall, a narrow strip of woodland, scarcely larger than one of our own woods, whose lower boundary wriggled across the slope like a snake. The wood is called Serpentwalk Plantation. So perhaps it really is possible that some day March Wood and Spring Wood will be named on the OS 1:25,000 map.

South-west Scotland is well-sprinkled with place-names of diverse origin, as could be expected given the area's turbulent history. The earliest names derive from Brittonic, which apparently was similar to Welsh. Old English place-names appear after the mid-7th century, when Galloway was conquered by the Kingdom of Northumbria. Gaelic names are commonplace, are frequently anglicised, and there are straightforward English names intersown across the area. Curiously, Scandinavian influence seems to be quite sparse, yet just a few miles across the Solway,

Cumbria is littered with place-names ending -by and -ham and -thwaite.

Along the Solway coast are names that refer to the wildlife associated with the place. Partan Craig, in Orchardtown Bay south of Dalbeattie, may well have been named such after the edible crabs (*partan* in Gaelic) that would have been found along this rocky shore. And not far from New Abbey, on the Nith shoreline, Craiglebbock Rock is believed to be named for the flounder fishing around it, *leabag* being a flounder in Gaelic, so Craiglebbock would derive from *creag leabag*, flounder rock.

And then there is the intriguingly-named Port O'Warren Bay. In past centuries this was a favourite landing-place for contraband, and the smugglers would have used horse-power to carry the goods up the steep incline behind the bay. It is thought that 'warren' may well be a corruption of 'garron', those sturdy sure-footed ponies that have also lent themselves to another placename that does not yet appear on any official map.

Eight: Caravan.

Close to midnight one Saturday evening, I left myself a memo on my pocket voice recorder: "On this day, the sixteenth of April two thousand and eleven, I have, possibly inadvisedly, suggested to Karen that I may be prepared to live in a *caravan* with her at White Croft Cottage. I may live to regret this, but I hereby do declare that I said this, in the presence of Sky the dog, who is the witness, and Karen, the other party, who is therefore not a witness. As she has generously pointed out, Sky is asleep, so this possibly means nothing..." Drink may have been taken, the words a little slurred, but the promise was made, however reluctantly.

Up till that rash evening I had not been in favour, as I knew that living in a caravan, just like living in a boat or a hut or a tent, requires a higher-than-normal degree of orderliness. And I had shared house-space with Karen for long enough to know that this would be, as they say, a challenge.

However, we had come to realise that the to and fro between Paradise and Garronfield was expensive in time and fuel. It was a round trip of 60 miles, an hour and a half of travelling. To be on site would therefore save on fuel, travel time, and we wouldn't have to pay rent. The attractions of the prospect had been steadily growing in me, and there was something else, something that resonated deep in my core.

I wasn't exactly born in a caravan, but I lived in one for the first three years of my life. Or, rather, more than one, as I can tell from the archive of family photos I inherited from my mother. My parents started married life living in a small touring van. As my father's job prospects improved they moved up to a 22-footer, and I spent my first few months of life in this. They then bought a 28-footer, so that I could have a room for myself. This was the mid-1950s, and no doubt there was still a shortage of housing after the extensive damage of the blitz, coupled with a post-war increase in births.

I had an Aunt Polly, not a blood relation, who lived in a pre-fab somewhere in Birmingham. I recall occasional visits to her home, one of scores of pre-cast reinforced concrete structures in a large estate, hurriedly created to meet a chronic housing shortage, and supposedly designed to last for ten years. The aesthetics of the pre-fab were none of my concern at the time, and to me it looked like a normal little cottage, with windows on either side of the front door. Aunt Polly collected three-penny bits, keeping them in a jam-jar, which she would present to me when we visited. The hoard of heavy coins was like treasure.

It is almost certain that my parents would have chosen to live in the country, rather than an urban estate, and residential caravan sites were the alternative to pre-fab housing. My own memories of caravan life, needless to say, are tenuous. If I can recall anything, it would be a sense of warmth and safety, and an underlying happiness. And if a further spell of caravan life could recreate that happy, safe warmth, then it was worth a risk.

Twelve months later our caravan arrived, a 36ft by 12ft static, bought from a couple who had finished renovating their home, and had used the van as a site office, loo, and kitchen more than anything else. It was twenty years old, and the décor was that far out of date. But we intended to strip most of the furnishings anyway, so that we could put our own stuff inside.

In the intervening time we had gone about providing ourselves with the necessary services: water, electricity, phone. Scottish Power and BT connected the last two, but we did most of the work, digging trenches for ducts that would carry the cables, and building a temporary weatherproof cupboard for the fuseboxes and sockets and meters. We discovered a strange phenomenon: whenever we opened a trench, within hours it would begin to rain and flood the excavation. We became adept at pumping muddy water, handling subsoil that was the consistency of porridge, and getting plastic ducts - that wanted to float - to stay in the bottom of the trench.

We were two miles from the nearest water main, so we had to provide our own supply. Here there was a continuum with the past, because the cottages had almost certainly been supplied, in their first decades, by a spring a short distance north-west of the buildings, which was bubbling away quite energetically as we carried out our excavations. We'd intended, once the cables were in place, to cap the spring by burying a concrete drain ring over it, on a bed of washed gravel to act as a filter, and pumping the water to where the caravan would be sited. Later on the water would be pumped into the cottage.

As I dug the trench for the electric cable towards the gable end of the buildings I achieved something that had not happened in a century and a half. I moved the spring. Water suddenly bubbled up in the trench, and within minutes the traditional site of the spring, a few yards away, had dried up. This was serious water, not like surface rainwater that would be intermittent and temporary, but a constantly-flowing source. The trench was full to the brim within an hour.

I had to dig a second trench, at right angles to the first, leading to lower ground, in order to de-water the main one. The spring continued to bubble up, and even when I backfilled the trench it did not return to its original place, despite much compacting of the backfill with the digger bucket.

The new site for the spring was not the best, not least because it was directly in line with the electric cable. No problem, I said to myself. If I've moved the spring once, I can do it again. So we marked out the ideal site, just a short distance from both the original source and the new one, and I began to dig. By the time I had gone down about seven feet, and not one drop of water had found its way into the hole, I realised the spring was staying put in its new location.

Eventually we did get it capped, although because we could not put the concrete ring directly over the upwelling, because of the cable duct, as much water emerged outside the ring as flowed into it. Nevertheless, we had a decent reservoir of clean water, filtered through gravel, that would meet our needs. All was now in place to service the caravan.

It took us two days to winch the van into place, beyond the far end of the cottages, where it would be out of the way of construction, and would have a pleasant view from the living-room's bay window, down the valley and across the Solway to the Cumbrian fells. This was the end of August, and our tenancy of Paradise Cottage would be up for renewal in mid-November. We decided we would let Paradise go, and be in the caravan just in time for the onset of winter, as several of our friends helpfully and encouragingly pointed out.

It was a lot of work, but we managed it. The interior was gutted to accommodate our own furniture, the kitchen revamped, a log-burner installed in the living-room, the underside of the floor insulated and sealed to prevent draughts, a decking ramp and platform constructed to the two doors, and our shed at Paradise was dismantled and re-erected alongside the caravan, to serve as wash-house, outside toilet, and store-room.

We moved in stages over the course of a week, fetching stuff in our horse-box, which, on the second day, moved Finn and Mister to their new home. The final load was brought on November 11th: our main furniture shoehorned into place, the stove installed and tested, and we collapsed into our chairs at the end of a full day - so full that for the first time in many years I forgot about The Silence on Armistice Day.

Although we told ourselves we'd only be in the van for a couple of years, we set about making it as home-like as possible. Inside was cosy, with books on shelves and some of our 'nice things' on display. Outside we created a garden: I cut an area of coarse grass until it became a lawn, we made flower-beds, and a sitting-out area with table and chairs. The bird-feeders went up near the outhouse, and I put out peanuts, seed, and scraps. There were no takers for a couple of days, then a pair of blue tits arrived. Before long we had a regular clientele of common garden birds, together with more unusual species - reed bunting, stonechat. In fact these last were the natural inhabitants of the place: we had intruded our domesticity into their wild territory. And as it turned out, we had inadvertently sited the caravan across a badger track. Once they were active again in the early spring it was a common occurrence to be confronted by a disgruntled badger, rootling around the base of the bird feeder, eyes glaring white in the torchlight, only moving away if pressed too hard.

If we were warm and well-lit and cosy inside, we were still aware of how close to the outside we lived. Behind the thin plywood walls was a wooden framework that was quilted with inch-thick rockwool insulation. Beyond that, the aluminium skin of the caravan, and that was all. Without the stove I doubt we would have stuck it, that first winter. I slept with a fleecy hat over my ears, and thick socks to keep my toes warm. But in the living-room and the kitchen and dining-room, which were open-plan, the woodburner did its job – sometimes so well that we had to open the living-room door to let some heat out.

And we became so aware of the sound of the weather, especially rain. Even drizzle sounded on the roof, a shsssh. At the start of a shower there would be the staccato of the first big drops, and the downpour to follow might easily overwhelm conversation or even music on the radio.

Then there was the wind. Although I had anchored the caravan well at each corner and several points in between, our first gale was experienced with some trepidation. The van did not move, but it trembled with the heavy gusts, the whole structure flexing ever so slightly in the tumult of pressures. And one night just before that first Christmas, as a gale wound itself up towards storm strength, the roar of a big gust hurtled towards us up the valley. When it struck the end of the caravan the big bay window smashed into four pieces and fell into the living room, on to the sofa, where neither of us was sitting.

Karen had been across to the outhouse, and had just opened the caravan door when the blast struck. I heard her shouting that she couldn't get the door closed, and I shouted back that I'd got my own problems to deal with. I was standing in the middle of the floor, being lashed by horizontal rain that was hissing and snapping against the stove, staring into the black teeth of the storm, and wondering what the hell to do.

I switched off vulnerable electrical equipment and scurried down to help Karen with the door. We got it locked securely, then my head went into emergency mode. Fortunately the rain had been just a flurry, so at least for the moment our stuff wasn't getting soaked. But we needed to plug that gaping hole, six foot by three foot. A sheet of plywood was needed, and we had some. Waterproofs on, tools gathered, and in the brief respite between gusts we hauled a sheet of ply into the back of my pickup, as I knew we wouldn't stand a chance trying to carry it in that wind. Karen lay on it to stop it blowing away, and I drove the short distance down to the caravan. A pocketful of screws and the power screwdriver. The wind actually helped us, once we had struggled the sheet into place, by holding it pressed against the window-opening. My relief when the first screw took firm hold, through the ply and into the aluminium skin, was immense.

Soon we had the sheet anchored, and the emergency was past. A slightly draughtier living-room than earlier that evening, but we plugged the gaps with rags and teatowels, and the stove soon did its work.

In the calm of the following day I took down the ply sheet and cut it to fit inside the window-frame. Sealing around the edge with silicone made it draught-free. As it was just before Christmas we couldn't get a replacement window until the new year, so we drew the curtains across the plywood and lived without the view for three weeks or so.

That winter gave us a demonstration of all weather types. In mid-January we had our first snowfall, a light dusting that lasted little more than a day. At the beginning of February there was a heavier fall that fully covered the ground, and left dollops of snow on gorse in bloom. The ponies had haynets to sustain them.

But it was after the end of meteorological winter that the season returned full force. In the last week of March it began to snow, late afternoon, and the wind picked up. It plastered the windows in the living-room with a thick layer that brought a premature dusk and early lights on. Through the night we heard the blizzard sighing, snow-laden. In the morning we woke to a startling scene.

Drifts were heaped around the caravan four feet high, but strangely the wind had swept a clear space around the caravan itself, so that it seemed we sat in a white bowl, or the crater of a cold volcano. As we ventured out we discovered that the hollow places had been filled, yet the high places kept clear. The deep gulley of the burn was full to the top, a snow-field hiding its very existence. When the fresh snow froze hard a day or two later, I could walk across the burn at any point, five feet or so above the muted trickle of water.

The track to the road was full to the brim between walls and had to be cleared with a snow-plough. Yet Paradise Meadow and Top Field were virtually clear, and the ponies were able to graze. As the first day went on a variety of birds not normally seen found this open ground and joined the garrons in feeding on whatever they could find in the as-yet unfrozen soil. So curlew, common gull, and lesser black-backed gull came to us for the first time.

After a couple of days the drifts around the caravan were sun-thawed on their upright faces, and came to resemble overlipping waves, seemingly about to crash around us, yet locked tight each freezing night as stars glittered coldly overhead, and ice-crystals glinted back at them. The thaw was slow and prolonged: there were drifts behind the dykes until mid-April.

Nine: Outside In.

I had thought there would be a gradual transition between my old life - before I knew I wanted my own wood - and the new life in which I longed to immerse myself. I'd imagined that I would cut down on the outside work over a length of time, and spend more and more of my efforts in restoring the building and looking after the land.

In the end the transition was almost abrupt. I completed the jobs I had in hand, but did nothing to seek new work, letting go of the contacts I had while we were at Paradise. I think that subconsciously I didn't want the dilemma of being offered work, and, since those who had employed me in the past no longer knew where I was, fresh offers did not come. It wasn't as though I needed to earn my keep: Karen's salary was enough for both of us to live on.

Then my trusty pickup failed its MOT disastrously, requiring a lot of money spending on it. As I no longer needed it for work, and supplies of materials for the building project were delivered to site, the decision was to retire the 20-year-old vehicle. It would stay, however, and live out the rest of its usefulness on our ten acres.

So all of a sudden, it seemed to me, the outside world shrank down to a shadow of what it had been. I was used to driving thirty miles or more to work: through the first winter after my pickup was off-road, the furthest I ventured was the five miles into Dalbeattie once a fortnight in Karen's jeep, to visit the library and fill petrol cans. The following summer, as work on the building intensified, my longest journey, for week after week, was to the end of the track to put out our bin, and collect it the next day.

Being self-employed tends to inculcate self-discipline. For almost forty years I had got by without having a boss, other than myself. Although there were stretches where I was obliged - by my own sense of responsibility - to turn up on site day after day, more often I had been free to organise my work to suit myself. There were rare mornings when I turned off the alarm and put my head under the covers until noon. Mostly I got up and got on with it.

In the previous decade I had learnt, more and more, to work with the weather. This was helped by broadening my business activities to include firewood and growing willow. These sidelines meant that I had work under cover for much of the year, and especially through winter, so that my waterproofs were rarely tested in the extreme conditions I had endured previously.

Now all that had changed. There was no-one expecting me to turn up to work for them, and I was more often than not alone on a muddy site with a caravan, a workshop, and a derelict building. I felt exposed, in more than a physical sense. I was left depending, more intensely than ever, on my own inner resources. I was embarking on a project for which I did not have much of the relevant experience. And a good deal of our stuff was stored outside, under tarpaulins to keep off the weather.

So we took delivery of a wagonload of poles, boards, plywood, and roofing sheets, and spent the best part of two months putting up three timber outbuildings, utterly functional but still aesthetically pleasing, and even more pleasing in being able to store materials and equipment under cover.

Then began a trawl through the lexicon of building: scarcements, sarking, u-values, dwangs. People speak of steep learning curves, but ours was almost a vertical line. Frequently we had to embark on a job we had never done before, and get it right first time. I did have some experience on the outskirts of the building industry, and a pretty good idea of what I wanted to achieve, but construction techniques had moved on considerably in recent years, and once I was outside the comfort zone of things I actually knew how to do, the work seemed perplexingly complicated. I was often terrified at the prospect of what I had to achieve.

That is not to say there was no satisfaction to be had: there was, and frequently enough a sense of exhilaration as things did go right, as the pieces of the big three-dimensional jigsaw puzzle came together, and the results looked good.

When every day is different, often challenging, and sometimes perplexing, it became vital for me to have a routine that would provide a framework to live within. I had listened often enough to others telling how, when they were building or renovating their house, they would usually still be working at midnight. Mindful that I was approaching sixty, still fit and strong, but wanting not to be worn down or made ill by the project, I determined to organise life so that I could survive the long haul.

After getting up I would do the domestic chores until breakfast. It was important to get these mundane tasks out of the way early, as they would be unlikely to get done later in the day, and then chaos would ensue. I am not someone who lives happily with chaos, my natural inclination is to bring order to confusion. So, first thing, wood for the stove would be brought in, stuff would be cleaned and tidied and put away, washing machine and dishwasher loaded and set going.

After breakfast, work on the building, or maintenance of the land if that was demanded. A break for lunch at half past twelve, then another stint until six o'clock. Sometimes, but it was rare, I would work on past six, though usually with a sense of diminishing returns. I do not make a religion of time management, but it does help me. I keep it simple: there are two ways of managing things, the first of which is to carry out a task for a set amount of time, then stop and move on. The second is to embark on a task and carry on with it until it is completed. The nature of the job often determines which way to manage things, but other times there is a choice. I tried to make sure that I didn't start something requiring completion in one go, unless there was adequate time to do so. It didn't always work.

One of the things that could make a mess of time management, and that became apparent early on, was the fact that almost everything would take three times longer than anticipated. I was glad that I wasn't having to quote for the work I was doing - I'd have been out of pocket, even bankrupt, quite soon into the project.

So I strove to maintain a semblance of structure to my days, fending off chaos, balancing extremes, tightrope-walking. As the connection to the outside world became more tenuous, I found that there wasn't anything I missed greatly, and discovered that instead I was increasingly content to be limited to my ten acres. In fact it was no limitation at all, since those ten acres consist of more than six million square inches, and to get to know every inch, as I intended, would be the work of a lifetime.

Ten: Neighbours.

To stand in Top Field and turn through all the degrees of the compass reveals little of human habitation and activity. There are farm buildings along the south-western skyline, but the farmhouse is further back and we are not overlooked from its windows.

One window does peer down at us, from a cottage high up on Hawthorn Hill to the south. The place is almost hidden by trees. No-one lives there all the time: the window is seldom lit.

Then there are the stone dykes, that separate field from field, and field from wood. In the fields are sheep and cattle, but those who tend them are seen only infrequently. Their houses are hidden behind ridges, beyond woods and shelterbelts. Whole days can pass without sight of another person in the landscape.

To the north, above the beeches of Auchenskeoch Wood, the twin summits of Round Fell and Maidenpap are prominent. The sun seems to dwell on them, lighting bold green conifers, purple heather, and orange bracken in season. In winter these hills will be dusted with our first glimpse of snow.

To the east, immediately behind the cottage, rises a small hill which is unnamed on the modern map, but was marked as Barn Hill on the 1854 Ordnance Survey map, so I will take the liberty of renaming it, although the hill is not mine. It is an outcrop of granite, and the rock pokes through the vegetation at the summit, which is crowned with gorse and a single rowan. The sun rises here in midsummer.

Barn Hill obscures further horizons, until its southern end drops away, and the corner of Criffel's granite massif forms a long and diminishing skyline. Directly south-east there is the cup of our valley, which gives us our longest vista. One field below Garronfield a fringe of trees traverses the valley, hiding what is immediately beyond. In amongst these trees one pine rises head and shoulders, and most of its upper torso, above the rest. It is a striking feature that draws the eye, and in clear conditions the eye, once drawn, can make out hills in the distance.

The nearest of these is Skiddaw, by Keswick in northern Cumbria. Other fells range out to the right of it, the last one visible being Helvellyn, by Thirlmere, more than 40 miles distant. In the clearest visibility many features of that landscape can be made out: familiar features, since they formed the horizon from the front of our cottage in Blencogo. The view from our back garden was across the Solway to Criffel, and the lower-lying ground to the west in which Garronfield is nestled. We had looked at that view almost daily for ten years, never imagining that eventually we would settle into it.

But from Garronfield that Lake District horizon is soon obscured by nearer heights: Millbank Hill, then Hawthorn Hill, and nearer still the tree-covered summit of Clonyard Wood, due south. Turning towards the west there is a steep field with the farm buildings at the top, and beyond, just visible above the nearer skyline, the forested top of Bainloch Hill.

The last segment of the compass is filled with the coniferous presence of Bennel Wood, where the sun sets for much of the year, where peregrine falcons nest, and where badgers have their sett. There is a fringe of large broadleaves, but the interior is packed with spruce, seasoned with clusters of larch. The top end of the wood slopes to the road, and last of all is a vee of sky between the wood and Round Fell, towards which our track heads.

As far as I am aware, all land in Scotland is owned by somebody. The sheep and cattle that graze the pastures around Garronfield are evidence of such ownership. Barn Hill is part of a block of fields that was one lot in the sale of Auchenlosh estate. It was bought by Alan Carter, who farms at Tokirra, about five miles to the north-west. His livestock circle the hill all summer. The cattle are taken away in the autumn, the sheep remaining through the winter. He visits regularly, driving round the stock in his jeep, a wave if he sees me. His workers come to spread fertiliser from time to time, and also wave if they see me.

At the bottom end of our western boundary, our neighbours are the Drummonds at Castle Farm, due east of Garronfield on the main road, and their last field to the west abuts our place for a short length. Castle Farm is the historical seat of Auchenskeoch, and the castle ruins nestle among the farm buildings. For the main part of our western side the March Dyke marks the boundary between ourselves and the Kenyons at New Farm. Their steep field with the buildings at the top occupies the view from the cottage windows, and is regularly dotted with livestock, who come and go like tides throughout the seasons.

Bennel Wood begins a stone's throw away from our boundary, but feels like a neighbour as it is such a dominant presence in our landscape. But who the owners are is unclear: some sporadic work was done in the wood when we first had Garronfield, then nothing. The access track from the main road has become choked with gorse and ash saplings.

Where the track from Garronfield meets the main road, it does so between two properties. On the right is Auchenskeoch Lodge, which when we first knew it seemed to be continually on the market, but was eventually bought by Craig Drummond from Castle Farm, and his wife Fran. On the other side is Lodge Wood Cottage, which was an outbuilding of the lodge. It was derelict when Vic and Suzanne Dunlop bought it and the woodland behind, and set about transforming it into a fine house. These two were our first friendly contacts, and their kindness and generosity helped sustain our earliest efforts. Their borehole kept us supplied with water when our spring dried up for the first time. I frequently consulted Vic about building matters, and his engineering mind willingly grappled with complex problems. His ability to do mental arithmetic was, to me, astonishing. He would usually arrive at a figure before I had even found my calculator.

Our farming neighbours do not impose themselves on us: they get on with their own business and leave us to ours. When we meet, infrequently, they are always friendly and want to know how we are getting on. They have cattle and sheep, we have ponies and trees, so we don't tend to discuss the price of fatstock or the cost of fertiliser. Most often we see them at a distance going round their stock on quad bikes, or hear them shouting at unruly dogs as they round up their beasts.

In time our circle of friends has extended outwards from our immediate neighbours, and again we have found kindness and generosity, almost without exception. It has become clear that to live in such a scattered community produces, in so many, a sense of self-reliance coupled with a deep respect for, and toleration of, those living just down the lane, or up the hill, or round the corner.

Eleven: Sanctuary.

Coming back to Garronfield after any visit to the outside, there is always a sense of reaching a safe haven. For the most part it is a peaceful place, but it is not always quiet. We have disturbed the peace ourselves, often enough, during the phases of building activity. But when the digger is shut down, the mixer falls silent, and the dumper is parked up, quietness seeps back like mist.

When we are not noisy the sounds of outside become more evident. At times the slope of Alan Carter's field behind the cottage is crowded with animals, continually grazing, tearing at the summer grass with eager tongues, and a soft sound carries down the hill on still days. Cattle compete with each other to scratch noisily at the straining-post at the top of Orchard Bank. The bull roams the fence-line, his dirty white hide tight over hillocks of muscle. He groans constantly as though in pain. Perhaps he is.

On the other side, New Farm wakes early. They start milking at around four o'clock: by six the sound of machines reaches down if the wind is in that direction. We see only the upper parts of their buildings, so what takes place at ground level is invisible to us, and we are unaware of farming activity until it makes a noise. Only when one of them on the quad bike comes through the top gateway into their steep field do we see any movement, other than the rhythmic circling of their sheep and cattle.

Sometimes we wake - Sunday mornings seem to be a favourite time - to the visceral thump of shells and stuttering gunfire. I always think this is how the day begins for all too many people in the world. Here it is harmless enough: the military playing war-games up at their practise range at Edingham, known as 'The Admiralty' locally, as torpedoes were manufactured there during World War Two. The place is studded with concrete bunkers and dark entrances into underground chambers.

When the military are on manoeuvres at Edingham the peace can be further shattered by low-flying helicopters of various sizes. Most of them seem to use our valley to sneak along, perhaps below radar. At first the ponies spooked, but after a couple of scares they don't even break off their grazing. Once two helicopters passed over, east to west, with cargo nets slung underneath. The first net held some kind of small armoured vehicle, the second held an artillery piece. I put my faith in someone being good at tying knots, as they flew over the bottom end of Garronfield and, it seemed, directly over New Farm.

Such noisy, low-level incursions are rare: two or three times a year. But other aircraft pass over daily, flying at thirty thousand feet, their noise a low mumble when it is calm. Any sort of breeze blows the sound away, and the planes pass over unnoticed, unless on clear days one catches the eye, silvered by sunlight, unzipping its own white track through the upper atmosphere.

And there are the occasional irregular sounds from outside: a chainsaw whining in the distance; a noisy car up on the road, or an ambulance siren; some sort of machine growling away, hidden in the depths of Clonyard Wood. A few times, when it falls still, I hear a soft roaring coming from the south. It is similar to the sound of the breeze in the top of the wood, but the trees are motionless. The first couple of times this happened I checked the tide tables and it seemed to coincide with either the incoming or outgoing tide.

I imagined that what I could hear was the cumulative whispering of miles of tide-line, fingering or dragging away from the mudbanks, slapping and sizzling around rocks, rushing into pools: the sound funnelled along the valley and amplified in my willing ear. I spoke this thought out loud at a party one evening, and was told that most likely what I was hearing was the roar of the massed ranks of wind turbines offshore in the Solway at Robin Rigg. I pointed out that this happened when the weather was calm, but my informant was undeterred: there is always a breeze out at sea, he said.

What is beyond doubt is that the tide brings mist to us. On a July evening, with high tide two hours away, the invasion begins. It seems sudden, as I haven't been aware of the first incursion. When I notice, the valley below us is already full to the brim: Hawthorn Hill and Clonyard Wood have vanished, and Alan Carter's field is quickly disappearing. The cattle are noisy, calling out for each other in the murk. Watching from the caravan, our trees become silhouetted against the white tide, until they fade and are overcome. It is absolutely still.

For a time the cattle become quiet, as presumably mothers and offspring have found each other. Then the bull starts roaring again, probably left behind by the nimbler herd. For more than an hour the mist creeps slowly towards the caravan and finally engulfs it. I think of what this would look like from above – in all probability the higher ground on either side would be mist-free, leaving just a narrow white tongue probing its way inland from the coast.

Then, over the space of half an hour, the mist retreats, shrinking to little more than a lens in the hollow below the bottom boundary. There is a full moon in a clear sky, and it seems as though the moonlight has pooled in a pale fire under the remnant of mist. It is a very mild night, completely still, and at last the bull has stopped calling. The silence is absolute.

There is a meteorological distinction between mist and fog, and it is down to visibility. If objects 2000 metres away are obscured, that is mist: if objects at 1000 metres (or less) are obscured, that is fog. What I have described is evidently fog, not mist, but I have a problem with this. I associate fog with the sea and with cities. It is a dense, heavy, inert thing that could linger for hours. To me, mist is altogether more nimble: it steals in and drifts away, sometimes returning for another foray. It is what we have at Garronfield.

I am at odds with the scientific definition, but unrepentant. Perhaps my recent habit of naming things here has caused me to overstep the mark, leading to an interpretation of natural phenomena to suit myself. But I am not trying to redefine the weather, just to blur the hard edge of science a little, as mist blurs the distinction between fence and tree, between earth and sky.

And on still days, when I hear that soft roaring to the south of us, as though a breeze is among the tree-tops of Clonyard Wood, though there is no breeze, I will think of it to be the tide, incoming or outgoing, rubbing itself against the edge of land. It is the sound of the sea, not the wind turbines.

Whether the air is still, or turbulent with winds, there is one particular place that has its own aura of sanctuary. It is the bottom-most corner, where the singing started, furthest away from the road, sunk down by the burn and sheltered by trees. Sky the dog is buried there.

Brambles grow in profusion along the line of the stone dyke, and one September Suzanne came to pick blackberries. She spent the best part of an hour filling two bags with ripe fruit. On her way back she stopped to chat. "You know, it's so peaceful down there", she said. It was the last time she was well enough to walk down to Garronfield.

Twelve: The Imperative of Spring.

The land breeds stone. Large boulders surface like blind whales, smaller ones are eerily skull-like in poor light. It would seem they have newly appeared, but it is simply that the thin veneer of turf and moss has sloughed off. Some bear scars from their journey, as they were scraped southeastwards in the underbelly of ice. Always there is a paring away to the bone.

Torrential rain engorges the watercourses and at the end they are reshaped - not major changes, but subtle rearrangements. The hard clay glacial till that underlies us, that forms the bed of the burn and the sikes, is washed into multicoloured slews of gravel - grey, white, rust-red, gold. In strong light the stones glitter with crystals.

And out of the land the Galloway cottages appear, like low granite extrusions, tucked into shelter where shelter is possible, or stern-faced, confronting the south-westerlies and stinging rains. Their eaves do not oversail by more than a couple of inches: stone copings at the gables weight down slates against winter gales. They are modest, they do not stand tall: they hunker down to withstand the tide of years.

In contrast the green life of the place rises to greet us.

'You do not own the land: the land owns you'. In my head I hear those words spoken - or rather, sung - in an Irish accent. I'm certain it is a line from a song, though which song, and who the singer, I can't bring to mind. But whatever truth there is in that saying I learn anew each year, especially in Spring.

It always starts slowly, as the grip of winter becomes tenuous. There is a quiet time when last year's vegetation lies shrivelled and beaten flat, when you can see through bare-branched woods, and animal life seems to be lying low.

Then a first primrose appears on the bank of the burn, across from the Singing Place. Robins chase around the outbuildings, defending territories. Lesser celandines begin to spangle the rough grass along the edge of March Wood. These signs are the first whispers of a flood that is about to overwhelm the place, but every year we welcome them, not thinking about the unstoppable surge that will engage our energies in the months ahead.

It would have been easy to let the land please itself, to let it have its explosive way, erupting in vigorous abundance from every nook and cranny. But we want to enjoy the place ourselves, as much as possible, and that means keeping open pathways we can easily walk along. To this end I scythe yard-wide paths with the brushcutter, some following the straight lines dictated by tree-planting, others broadening the narrow informal tracks we had already begun to make as we moved around the place.

These informal routes, or *desire lines* as they are known, are indicative of the way humans move around their particular terrain. Planners and landscape architects have become fully aware that people generally make shortcuts from one place to another, and 'keep off the grass' signs are likely to be complied with only by those who are exceptionally conscientious, or not short of time, or both. Designed landscapes commonly have hard or soft physical barriers to shepherd people through them.

Today's maps can reveal ancient and historical desire lines, indelibly printed on the landscape. Some of the most ancient British tracks are the ridgeways, dating back at least to the Iron Age. They follow generally direct lines across downland and moors, open ground emerging from a tree-covered landscape. A modern road atlas has page after page criss-crossed with the red lines of A-roads, many of which faithfully reproduce the web of highways built during the Roman occupation. They tended to take a straight line between places, in order to move the legions around the country as quickly as possible. Similarly, in Scotland eighteenth century military roads scythed through the terrain, avoiding habitation and obvious ambush points. Drove roads also followed a generally straight route, to shorten the distance that stock had to travel to their markets.

At Garronfield there is a mix of straight and meandering paths. The straight paths mostly follow lines made by the steel tracks of the 13-ton digger, and have become narrow green tree-lined avenues. The meandering paths wander among the trees, taking the easiest course between them. A sinuous path leads up from Irish Bridge, along a curving avenue of rowans, and into the Holly Grove. From there one track heads to a plank bridge across the burn, and another weaves among the trees at the top end of March Wood.

This path makes an abrupt detour near the wood's topmost corner. Here, when I was opening the path for the first time, in late May, I was watching out for young trees, already submerged in vegetation, to keep them from being damaged by the hissing blade of the brushcutter. As I approached the top corner I saw a glint just ahead of the cutter. For a moment I could not fathom what I was seeing, then comprehension struck and I jerked the brushcutter up and back out of the way.

The glint was an eye: the small dark eye of a roe deer fawn lying absolutely still - as its mother had told it to - despite the horrible whining, crashing thing that had come nearer and nearer. The fawn's head was furthest away from me, and the blade may well have passed over its rear end, pressed into a hollow among the rushes. If it had been lying the other way, it is quite likely I would have killed the little creature, or at least wounded it.

For a few moments I stood and marvelled at the exquisite markings that so perfectly camouflaged the fawn amid the vegetation, then I backtracked a little, and took a different course to avoid the place - a deviation that has become a permanent feature of the path. I crept back an hour or so later and the fawn was still there, alive, unscathed. Over the next few weeks we had the privilege to see the youngster, and its mother, as they sheltered in March Wood during the day. The next year the doe had twins, and the year after that she had triplets. After that she did not come back to Garronfield to drop her young.

In autumn and winter the pathways are kept open solely by the tread of our boots, as we go about the work of maintaining and encouraging the trees, or, conversely, cutting them down. It has become something of a ritual to have a walk-about after Sunday lunch, a leisurely amble along the straight and the meandering paths, taking stock of things. In spring and early summer our particular pleasure has been to see the growing of the trees we have planted. From knee-high to waist-high to shoulder-high, they have now overtopped us.

But once the first flush of spring has greened the place, it requires diligence and some effort to keep the pathways open. It is as though rushes and thistles want to reach across the narrow divide and clasp their neighbours in a tangled embrace. Left alone, the tall vegetation would annihilate each track in a matter of days. Every week in June and July the brushcutter scythes its way through overhanging vegetation, reopening a jungle pathway.

When we first had the place, the boggy areas were almost uniformly colonised by soft rush - an odd thistle here and there, but nothing else. A close look might have revealed some diversity of plantlife hidden among the coarse stems, but photographs from the first year show only a sea of rushes, the result of continuous grazing. But when, that first autumn, I did the preparation for tree-planting, turning square yards of vegetation under, exposing clean black soil, I was also unearthing a hoard of dormant seeds, waiting their moment to burst into green growth - a moment that had now arrived.

To begin with, the wealth of the seed bank was not too apparent. Through the first summer after planting, as I hand-weeded around the new trees, the dollops had greened-up, but with generally low-growing plants such as buttercups and plantains, and the occasional dock. In March Wood, outside the rushy area, we were pleased to find a few purple orchids nestling in the rough grass, and a straggle of common spotted orchids in one place on the bank of the burn. In Spring Wood, the more fertile of the two plantings due to the nutritional input from several springs as well as run-off from Alan Carter's field, docks and nettles were more evident.

But subsequent summers saw a burgeoning of diversity. Thistles were the first population explosion: thousands of pink and purple heads, taller than many of the young trees. The bottom end of Spring Wood was particularly spectacular, the second summer after planting. And then came the umbellifer-forest. In March Wood especially, hogweed and angelica massed white and pink umbrellas among the trees, a sudden thickening of the wood, obscuring sight-lines.

As well as these tall, striking invaders (wind-blown seedings from elsewhere), a humbler flora began to flourish - cuckoo flower, ragged robin, stitchwort, betony, birds foot trefoil. Vetches twined among taller stems; gangs of foxgloves appeared in odd places; rare patches of bare ground sprouted those prehistoric bottle-brushes, horsetails. And purple orchids appeared among the young trees, though rarely in the same place two years running, a mystery that is as yet unsolved.

In the willow beds, late March or early April sees the imperative of spring take hold. The close-planted stools erupt with purple buds that quickly burst open to reveal the first green leaves of the season's growth. At around the same time the unwelcome squatters appear - thistles, nettles, docks, and willowherb. Although the willow is planted through a weed-suppressing membrane, these chancers manage to find lodgings in the slits in the membrane alongside the stools, and their vigorous first growth will soon smother the young willow shoots.

Willow is apically dominant: it wants to reach for the sky, and does not tolerate competition. It is a pioneer species, and was one of the first - along with birch and aspen - to recolonise Britain's terrain after the retreat of the last ice-sheet. My mollycoddled plants, in their sheltered environment, cannot thrive in competition with the coarse intruders, and so the imperative means a careful trawl through the willow beds to root out the unwanted. Failure to do so means the loss of some stools, or the dramatic suppression of growth. If the early purging is carried out with care, the willow's subsequent rampant growth will keep weeds at bay.

If we enjoy using our pathways, it soon became obvious that others do so also. One year in mid-January there was a fall of snow that ceased around midnight, leaving a couple of inches. The next morning we could clearly see how wild creatures had used our trackways through the night: roe deer before it had finished snowing, their slots part-filled, while badger and fox had turned out after the snow had stopped, their spoor clean and clear.

Thirteen: The Wood for the Trees.

At the beginning of *Wildwood, A Journey Through Trees*, Roger Deakin states that his aim in writing was to promote in people a greater appreciation of trees, so that rather than them just being seen in the plural, they might be noticed as individuals. Four years after our initial planting of March Wood and Spring Wood I began to experience an apparently contrary view: I started seeing the planted trees as *woods*.

I had not inherited a ready-grown wood, where I might well have spent time getting to know the individuals that made up that entity. Instead, where no trees previously grew, I had laboriously planted thousands of specimens, from a number of native families. They were my babies, and I looked after them as well as I could. As many as possible were provided with a mulch mat, a vole guard, and a shelter to protect them from deer. In the first growing season I spent long hours hand-weeding around them. I was full of pride when they thrived, and fretted when they looked sickly.

For the first two years there was not much impact: only willow attained any height. But in the third year the first individuals emerged from their shelters, which were later removed, and passed down to trees that had not had any initial protection other than a vole guard. Many of these individuals had suffered from browsing by deer, but they had their roots in the ground, and usually responded well to their sudden upgrade to first class.

That autumn, the end of the third growing season, I quartered the ground in both woods, seeking out the cinderellas that had been left to fend for themselves, and each time I found one I caught myself saying, out loud, "You *shall* go to the ball..." When it came to removing shelters from the more sturdy individuals, where they had outstretching branches I would clasp these together in one hand while tugging the shelter upwards with the other, at the same time saying (again, out loud), "Put your arms up!", as though to a toddler, while removing its jumper.

But it was midsummer of the fourth year we noticed, on our regular walk-abouts, usually after Sunday lunch, that there were places where we felt closed-in, where vistas were hidden behind the green cloak of trees that were overtopping us for the first time. "This really is starting to feel like a wood", I said - out loud, but to Karen this time.

Of course it was the burgeoning growth of the trees themselves that produced this effect, this sensation of wood-ness. As well as reaching for the sky they were reaching out towards each other, filling the empty spaces with their eager branches. In some spots it was actually becoming difficult to walk among the trees, and I was glad that I'd made the pathways, so that at least we could enjoy being in our woods without causing damage, either to the trees or to ourselves.

But there was more to it than simply tree-growth. What was happening overall was a transformation of habitat. At the point where the plantings were beginning to look like woods there was a balance between trees and other plants, especially where there was still open space between trees. Oak was our slowest-growing species, possibly not liking the peaty soil, and in the places where oak predominated, a diversity of other plants flourished. Where the trees were denser, beginning to dominate, to block out direct sunlight, there were the first signs of a change.

The transition had firstly been from open ground to what might be termed scrub, for although the planted trees would grow on, their early stages allowed for that diversity of plantlife to share the same ground. Where the trees were beginning to shade the ground, a subtle shift was taking place, with small areas becoming less diverse as shade-intolerant species failed to thrive. There was still a long way to go until a true broadleaf woodland habitat would become evident, but the process was starting to happen.

From the outset, the intention of planting our woods was to provide fuel, and this would be achieved by coppicing. The system of managing woodland by regular cutting actually replicates the transition from open ground to woodland through each coppice cycle. Where I have started to coppice small areas of well-grown trees, the opening-up of the ground layer produces an explosion of previously-suppressed plantlife, and the vigorous early growth from the coppice stools shoulders upwards through an abundance of nettles, docks, and thistles. By late summer the coppice stems have outstripped the rest, and are on their way to reassert dominance for the next few years, when the cycle will begin again.

Some trees are left to grow on, but only a handful in each acre. I do not want large-diameter timber: firewood of a hand's-width diameter is perfect to feed the biomass boiler that heats our home. Not only is timber of this size easier to fell and to transport, there is less work involved in the processing of it, as the logs produced do not need to be split. Coppicing is the way to produce the maximum amount of woodfuel from a given area.

And it is an old way. In his book *Ancient Woodland*, Oliver Rackham postulates that natural or self-coppicing ability in many tree species might actually have pre-dated human activity. Where huge mammals, now extinct, were able to topple and eat large trees, the evolutionary response of the vegetation was to put out fresh shoots, or suckers, to ensure survival. Prehistoric humans must have realised at some point that regrowth from a stump was more useful - more easily cut and managed - than the original tree. The evidence of coppicing in Neolithic times is found in the wooden trackways across the peat of the Somerset Levels, the oldest of which is dated to nearly 4000BC.

In the end, there is no contradiction of Deakin's assertion of the need to see trees as individuals. I do know my trees in that way, although I don't give them names other than the species names they are already graced with. At the same time I am aware of the gradual transition taking place, that leads towards woodland habitat. In this I have to listen to the trees themselves, to hear what they are saying about the place where they are planted.

Oak, as I have said, is doing least well of all the species introduced here. Having recognised this, I am steadily interplanting with coppiced willow rods, and clones of Scottish aspen. The hope is that the interplanting will act as nurse trees for the oaks, allowing them to delve deeper through the peaty soil to find the more suitable clay beneath. Then it may well be that oak begins to flourish.

Oak, ash, hazel. That was the very typical mixture we decided on at the planning stage, plus alder for the boggiest places. Ash has responded well to the conditions, growing away lustily from the outset. But the shadow of ash dieback hangs over us, and the first small signs of it have appeared as suddenly-dead leaves on one branch of an otherwise healthy-looking tree. We remove and burn these signs and hope for the best. If the best does not come about, and our ash trees die, there will be substantial gaps in our woods.

In thinking about this eventuality, and considering what to replant, it has been salutary to look carefully at what is thriving here, and hereabouts. We regularly drive along the B793 to and from Dalbeattie, and in those five miles pass several places where stands of alder, willow, and birch are predominant species, especially in Auchennines Moss, whose boggy ground echoes our own wet areas. Previously I had noticed this, without drawing a parallel, but now it is as though a woodland voice is speaking to me - maybe oak will thrive, maybe it won't; maybe ash will survive, probably it won't; but who is there to take their place?

Me, says the alder: I love it here. Sometimes I fall over, for no apparent reason, and sometimes I snap off at the base of my trunk without warning. But look how readily the broken stem puts out new purple-budded shoots, and how even when I lie prostrate my branches turn to grow straight upwards. I've got some strange habits, but here I can thrive.

Me, says the willow: plant me where you will, I shall grow. You only need to cut a rod in late winter, and stick it in the soft ground. Next time you look I will be a tree in leaf, already safe from browsing deer. I'm easy, I'm fast, I burn well.

Me, says the birch: this is my place. I'm difficult to begin with, and some you plant will die. But give me time, and give me grace, and in return I will give you gracefulness. You will look on me with pleasure as the years go by, and maybe my beauty will stay your hand as you come to me with your sharp-toothed saw. I carry thousand upon thousand seeds, and will sow them in the wind if you let me be.

Fourteen: Friends.

The friendly robin that follows me about in the winter is anything but. This bird, aggressive to its own kind - sometimes fatally so - feels no affection towards me, but is actually a bold opportunist, uniquely among our native birds. He (or she) balances the danger of close proximity against the potential reward of food. At some point in unremembered time robins discovered that certain human activities would uncover rich pickings of worms and insects. The one that shadows me as I dig the vegetable plot will dart in close to claim an unearthed worm, then moments later will not hesitate to drive off a rival with a stern 'tchick'.

But if the robin has no friendly feeling towards me, the opposite is not the case. I was a member of the Young Ornithologists Club from the age of eleven (the club badge was a silver and black hovering kestrel with the letters Y.O.C curving round beneath the bird's outspread tail) and even when I attempted to be a hippy dropout at fifteen and sixteen I was always aware of the birdlife around me.

Humans live in the same sort of world as birds do. Fascinating though the natural history of our native mammals might be, their activities are generally nocturnal, secretive, often underground: a world of night-time snufflings and barkings, of touch and smell. Although mammals are more like us than birds are, with visible ears and noses, and mouths with teeth, their lives, to most of us, are somehow *other*.

We share the bird-world of sight and sound. The more familiar birds we encounter in the daytime, and their behaviour is on display: we see them searching for food, hunting for prey, defending territories, fighting, courting, even mating. Only in the breeding season do the common birds become more secretive, hiding away on their nests. Even then there is a presence, as males sing lustily to proclaim their domain. And then there is the flush of fledglings, a population explosion for a few weeks.

Humans have song in common with many birds, though we sing for reasons other than territorial. Attracting a mate would be the closest equivalence. In unenlightened times songbirds were captured and put in cages - linnets, goldfinches, robins (all heaven in a rage). Nowadays we mostly just enjoy the singing of free wild creatures, and it speaks to our innermost beings, if we are open to it: the robin's winter song; the blackbird's fluting at dusk; the skylark's exultant cadenzas. Even birds that do not sing have calls that speak to our emotions: the first cuckoo in spring; the plaintive skirl of a curlew in the early morning.

Birds are welcome at Garronfield, and over time more than sixty species have availed themselves of our acres. The feeders outside the kitchen windows provide a close view of the commoner birds, and the occasional rarer visitor. The grey hoodie gang - jackdaws - calls by nearly every day to mop up breadcrusts. They seem nervy, springing upwards or sideways, squabbling. They size up the fare on offer, then try to fit as many crusts as possible in their beaks, often dropping one to substitute a larger bit. Then they fly off with their haul, to eat it somewhere in peace. The coal tit, by contrast, selects a single seed from the feeder and darts away to store it in some hidden cache. It is back again moments later, and the routine continues over and over. Sometimes it seems as though the bird never eats anything it takes, and apparently it is common for them to forget where they have stored their food, so in the end others get to eat it instead of them.

Once in a while a fierce-eyed blade scythes in from nowhere. The sparrowhawk targets the feeders, and if it fails to pluck a sparrow from the air, it then crashes into the elder tree in search of a victim, usually without success. The prey has already darted into the safe zone, a tangle of briers and nettles under the elder, too dense for the hawk. A blue tit opts for invisibility by hanging motionless on the peanut feeder, and is still there ten minutes later, long after the predator has left. When the danger is well past, feeding resumes gradually - chaffinches invariably the first to break cover. A single pied wagtail scuttles across to the feeders, and flicks about under them, driving off any competition. Even the brave chaffinches give way, and wait till the wagtail has had its fill and gone.

Birds are, for me, one vital point of contact with the entirety of the natural world. The robin reminds me that there is a hidden ecosystem of creatures all around me and underfoot. The bird is a portal into a world that is too often ignored, underappreciated. I don't begrudge the robin a few worms, but I am glad to discover that the ones that don't become tasty treats are there in the soil, working away, converting organic matter to humus, enriching the soil with nutrients, and improving soil structure.

As I harvest my willow crop, feet scuffing up the mulch of dead leaves on the surface, the robin dives in again and again to capture some minute grub or insect. It must be said that I do have reservations about the totality of insect activity in my willow beds, and I am not in favour of caterpillars eating the crop, but I recognise that my own activities are an insertion into the natural cycle of life. Even the small creatures we are not keen on play their part, being food sources for other creatures, and dealing with waste matter. It is said that without the likes of wasps and ants and beetles we would be up to our necks in dung.

And then there are midges. At Garronfield they are an occasional and tolerable menace, not so bad as those encountered further north in Scotland. I recall school camping trips to the west coast, before the widespread availability of effective insect-repellents, when midges drove me screaming into the sea in a futile attempt to be rid of the little biters. On a more adventurous school trip to Iceland, our party camped by Myvatn, a lake in the north of the island. On our first evening we were appalled to see the marshy ground along the shore seeming to float upwards towards us - millions of midges leaving their roost to feed in the long twilight. Fortunately for us they were non-biters, presumably vegetarian. They were a nuisance while cooking, getting into every bubbling pan. At first we attempted to exclude them - unsuccessfully - until we resorted to stirring them in, a little extra protein in our camp rations.

Here we splash on Jungle Formula and brave still summer evenings, sharing them with the incessant swooping flight of swallows, hoovering up midges by the beakful to stuff them into the wide gapes of their young in their nests. Later a pipistrelle or two will flitter about in the near dark, claiming another share of the midge population, which never seems to diminish, but we are glad that it can help to sustain bats and birds alike.

When we took possession of Garronfield in late June it was immediately apparent that many others were already comfortably in ownership of the place, and had been for some time. Everywhere we looked inside the buildings there were swallow's nests, mud cups stuck into wall corners, above doors, and on top of the electricity meters. Even as we walked around the adults were flying in and out of broken windows and holes in the roof, busily feeding their young. Probably upwards of a score of swallow generations had nested in the cottages since they were abandoned by humans, and the lineage had obviously bred a fierce tenacity in these birds that live in perpetual summer.

As we began demolition, their nesting places grew fewer. Nevertheless some were not deterred, and one pair nested inside one of the fireplaces, to-ing and fro-ing even when I was excavating the floor nearby with a digger, flying under the scissoring boom as though nothing was happening. As we began to rebuild, we inevitably provided a host of new locations for nesting - a new dry roof, open doorways and windows, and a framework of roof trusses with their inviting interstices. That April's arrivals could not have been more excited.

I set about excluding them, with mesh screens across all openings and along the under-eaves. Frequently they breached my defences, and I experienced alternate bouts of guilt - for keeping them from their ancestral breeding-place - and sheer exasperation at their doggedness. In the early spring I had installed four swallow nestboxes in the wood shed, and eventually the determined crew resigned themselves to using one or two of these, plus a number of other corners and ledges that they decided on for themselves. Over the next couple of years the swallows seemed to settle down into their new breeding-ground, although the first arrivals each year would check out the cottage initially, to see if I had overlooked a convenient place for them to set up house.

The arrival of the first swallow is always a notable event, and every year I am awed and humbled to think that the slender bird has made its way from somewhere in Africa to my home, in the hope that there will still be a nesting-place, and a mate soon to arrive. The departure of swallows is a different matter: we see them lining up along the telephone wire, restlessly chittering to one another, then they seem to have gone, but some still flitter about, catching a last meal or two before departure. Then there is a day when you know they really have left: it is an absence, an emptiness, a tinge of melancholy. The year I put up nestboxes for them, our birds seemed to be gone around mid-September. After a few days of swallow-less skies, a pair suddenly appeared among the outbuildings, racing around, calling excitedly to one another, briefly hovering at the nestboxes and the other nesting places, then away.

Amongst the other creatures that share our space, the red squirrel is most welcome. This is not because of some hierarchy of mammals, but in view of the well-known threat to the red's populations due to competition and disease transfer from the non-native grey squirrel. Greys are not far away: a few miles for certain, and possibly even on our doorstep, undetected as yet.

From the outset it was clear that red squirrels used the march dyke as a highway between Bennel Wood and Clonyard. The northern end is a switchback of gaps and standing sections, but further down the dyke is still in good condition, just a few topstones missing here and there. One morning I was at work among our newly-planted trees, alongside the better part of the dyke, when I spotted a squirrel heading at some speed towards me from the direction of Bennel Wood. Although I stood still, it spotted me when it got close, and sat up on its tail to assess the situation.

I felt as though I could see the cogs of its brain whirring. Continuing along the top of the dyke was out of the question, and turning back was ruled out as well. Eventually the clear choice was to descend to ground level, get past me, then return to the dyke top and continue on its way. I was astonished and amused to watch as the squirrel then began to climb down - on my side of the dyke. Halfway down it stopped, one eye fixed on me, as evidently the cogs were in motion once more.

Moments passed in this freeze-frame fashion until it worked out what was wrong, turned back to the top of the dyke and out of sight on the other side. It was not long before the squirrel regained the top of the dyke, a suitable distance beyond me, and continued on its nervy stop-start journey towards Clonyard.

Five years on from this incident we were no longer sure whether the march dyke highway was still in use, because it was masked from view by tree growth. We'd heard that red squirrels had disappeared from nearby locations, and could only hope that our commuters were still there.

One October lunchtime Karen and I were sitting in the kitchen when a red squirrel appeared on the ground from the direction of Spring Wood. It eyed the peanut feeder with interest, but then climbed into the nearby elder tree, where it stayed for some time before disappearing. I felt a great sense of pride that the squirrel had been able to traverse Garronfield under cover of trees.

Fifteen: Hinterland.

White Croft Cottage does not quite turn its back, but turns its shoulder to the north, a cold shoulder that sunshine cannot reach, even in midsummer. It is northwards, broadly speaking, where most of life beyond Garronfield takes place. Up there is Karen's work, as well as provisions, doctor, dentist, library, shoe-shops, ironmongers. I go up the track and turn left for Dalbeattie and Castle Douglas, right for Dumfries. Beyond the towns lies our hinterland, a northland of hills and lochs and forests.

This hinterland could be defined as the area between two rivers: the Dee in the west, the Nith in the east. A clockwise tour would begin at Kirkcudbright, where the River Dee flows into the Solway. A short distance up from the mouth of the river is Tongland Hydro-Electric Power Station, with its dam less than a mile further on, and the narrow Tongland Loch behind it.

The river catchment, consisting of the Dee, Loch Ken, the Water of Ken, and the Water of Deuch, was utilised in the 1930s for an extensive hydro-electric scheme to provide electricity throughout south-west Scotland. This entailed the building of seven dams, six power stations, and the driving of two tunnels to incorporate the waters of Clatteringshaws Loch to the west of the catchment, and Loch Doon to the north. At the time of its completion in 1938 it was the largest hydro scheme in Britain.

On the A713 Ayr road, beyond Carsphairn the land opens out and there is a first glimpse of a new era of industrialisation in the countryside: wind turbines, for which the uplands of Galloway are so suited, being high, accessible terrain in the path of the prevailing wind. The turbine blades flicker along the skyline to the north-west: to the north-east, clustered around the aptly-named Windy Standard, are many more turbines. There are a lot of them, and more planned. To be fair, few of them are visible from the main road, and the casual motorist would be unaware of their extent.

Looked at individually, turbines are graceful structures, unlike the grey pylons striding through swathes of clear-fell parallel to the road. But are there now too many of them? Unfortunately there is no let-up in our increasing demand for energy, and as coal is no longer acceptable as a means of generation, other means must be found. There is no limit to the amount of wind farms the government would like to see in Scotland. The stark truth is that so long as our demand for energy increases, there will be more wind turbines among our hills and other exposed places, which by definition are highly visible within the countryside.

The sign for Dalmellington proclaims it to be 'a village in the stars' - presumably because of the proximity of the Galloway Forest Dark Sky Park, and the Scottish Dark Sky Observatory near Loch Doon. Strangely, when I passed through the village on a bright afternoon in mid-June, all the streetlights were on. Perhaps the inhabitants light their streets by day, then darken them at night, out of respect for the Park's star-filled sky.

Heading north-east towards the source of the Nith, the road winds among the tell-tale bings of coal-mining, once an important industry in the area, now reduced to the remnants of open-cast. At Nith Lodge, where the road crosses the river that has its source on the western flank of Enoch Hill, the stream is little more than a couple of yards wide. It heads north, before turning east for several miles, then tending ever more southwards towards Dumfries.

Below New Cumnock the river has already attained a decent size. The A76 Kilmarnock to Dumfries road shadows its course, threading through sprawling linear towns, Kirkconnel and Sanquhar. In places the road is sandwiched between the river and the railway that also follows the long valley southwards.

Six miles above Dumfries the Nith flows past Ellisland Farm, where Robert Burns lived and farmed between 1788 and 1791. The farm is perched at the top of a bank close to the river, which here is wide and shallow, and chatters over its stony bed. The place proved to be unproductive - 'ruinous', as Burns himself put it - but during his stay his muse was not silenced. About one quarter of his total output was written at Ellisland, including *Tam o'Shanter* - apparently composed over the course of a single day, while walking and sitting on the river-bank - *Auld Lang Syne*, and *Parting Song to Clarinda*. The latter, better known as *Ae Fond Kiss*, is one of a number of poems written during his unfulfilled love affair with Agnes ('Nancy') Maclehose:

Ae fond kiss, and then we sever;

Ae farewell, and then forever!

The anguish of parting, of separated friends, is a common theme in Burns' poems, and underlies the whole of Auld Lang Syne. The first verse and chorus are most often sung in a rollicking, rumbustious fashion on New Year's Eve, but as anyone who has heard the song through to its end will know, there is an air of melancholy about it:

We twa hae paidl'd in the burn,

Frae morning sun till dine;

But seas between us braid hae roar'd

Sin' auld lang syne.

Eddi Reader's rendering of the song, on her album *Eddi Reader Sings the Songs of Robert Burns*, is to my mind the closest in tone to the meaning of Burns' words.

Above Dumfries the Nith contorts itself through a series of meanders, before slipping through town, past the flood-prone Whitesands area, and into the Solway below Glencaple, completing a skewed horseshoe-shape of a journey around the edge of our hinterland.

Between the Dee and the Nith is still, for me, mostly unexplored territory. We have visited a few places, briefly, and out of it come travellers' tales, and the music of names: Cloud Hill, Scaur Water, Windy Standard. There are sculptures to find and bluebell woods to enjoy. Now that we have just one vehicle between us, when Karen is working I am no longer able to jump in the driver's seat, turn the key, and head off. I do not resent my self-imposed exile in the least, but I am reading the OS Explorer maps with sharpened interest.

Sixteen: Keeper.

Once upon a time in Blencogo I became the proud owner of a fairly large and conspicuous sycamore tree. I was pleased to have the tree in my garden, and I looked after it: I cut out some dead wood in the canopy, trimmed back branches on the lower trunk, and made a terrace within its shade, where we could sit out on summer days, and in the evenings watch the sun sink among the Galloway hills. Before a storm I would usually go up and touch the tree's scaly bark, and wish it well.

But could I truly be called the owner of that tree, any more than I could claim possession of the chaffinches that made use of its shelter, the blue tits in their nestbox, or the carrion crows that built their nest high in its canopy one year? Rather let it be said that I owned a plot of land in which grew a sycamore, in the corner of the boundary between our back garden and Ridley's field.

If I was ambivalent about ownership of the sycamore, I am less so about the trees that are planted here. As I have already shown, I prepared the ground for them, dug their roots into the black soil, provided them with first-class shelters. I have tended them in their first growing, and have been alert to their condition as they have thrust themselves skywards.

However, if they are mine, they are not possessions. They are not like the binoculars that sit on the windowsill of my study, or the notebook open beside my computer, or the computer itself. Those are possessions: the trees are more akin to children. I can say they are mine, but in a real sense I do not own them.

The woods will outlive me: ownership of the land will pass to someone else. So what is the relationship between myself and the trees? The word *guardian* springs to mind, and has some resonance, although I am uneasy about the insinuation that I might be able to shield the trees against all misfortune - for I can hardly avert ash dieback, or temper a gale to prevent damage.

The term *curator* is much overused nowadays, yet in many ways it fits the bill: someone who has the charge of anything, and I especially like the OED's slant - someone appointed as a guardian of a minor, or a lunatic. *Steward* is another well-used term, as in agricultural stewardship schemes and so on, and yet again it goes a long way towards describing my relationship to the trees, although I dislike the connotation of officialdom, of being the servant of someone of higher status.

In searching through definitions to find the most suitable appellation, one synonym for all of the above has come to the fore: the simple word, *keeper*. In original use, one who has charge, care or oversight over any person or thing, especially charge of a forest, woods, or grounds. The term *gamekeeper* is an extension of this usage, and I am happy to be such - not that I will be culling crows or jays, or artificially rearing pheasants for shooting, but rather intent on keeping the diversity of wildlife here as best I can.

And I am delighted with the spin-off definitions of 'keeper': one who observes or keeps a promise; one who keeps a mistress (obsolete); a loop securing the end of a buckled strap; a bar of soft iron placed across the poles of a horseshoe magnet, to prevent loss of power; a ring that keeps another (esp. the wedding-ring) on the finger.

That's it, then: I am a keeper. A keeper of trees. A *tree keeper* - though that is awkwardly close to *treecreeper*, a bird that I would be pleased to spot at Garronfield. Well, never mind the awkwardness, tree keeper it is. Shall I be vain and award myself capitals? Perhaps not.

The beauty of it is, the word 'keeper' allows the sense of possession, of ownership, to lie beneath the surface of meaning without imposing itself. This is the perfect fit: the subtle melding of that possessive sense with the acknowledgement that the woods are being cared for and kept through the rest of my lifespan, to be handed on to the next keeper.

Seventeen: Hearth.

Sunday the 29th of January 2017 was a pivotal day in the rebuilding of White Croft Cottage: the place began to regain its heart. In the afternoon a sandstone slab, more than four feet wide, two and a half feet deep, and three inches thick, was walked up a plank through the kitchen doors and trundled on rollers into the living-room. There it was carefully eased into position as the hearthstone of our main fireplace. At the end of that day's work I poured us wee drams of single malt, and we first spilled out a small libation on the stone, before swallowing the rest.

I spent several days building the surround, stage by stage from sandstone flags sawn and dressed to shape, and we acquired an eighteen-inch dog grate to sit in splendour on the hearthstone. For this was to be my indulgence, my luxury, an open fire. We had installed a highly efficient underfloor heating system throughout the cottage, powered by a log-burning boiler. To continue the theme of efficiency would have entailed putting a log-burning stove in the fireplace, but there was one problem - me.

I'd faithfully served the stove we'd had in the caravan for four and a half years. It wasn't efficient enough to stay in overnight, but it was very effective at heating the van. The stove had done its job well, and each evening I could see the glow of burning logs and sometimes hear the crackle as the flames consumed them, but once the stove was filled, and the door shut, that was it until the next refuelling. My problem was that I itched to be able to poke and prod the fire, to encourage it, to help it fulfil its potential. But I was for the most part redundant: a stoker rather than a firemaker. I didn't want a glow in a tin can - I wanted a campfire.

There is no evidence to show exactly how or when or where our early ancestors first learned to use and then control fire, but at some point they did just that, undoubtedly taking flame from a wildfire as the first step of a long process.

It is difficult for us, accustomed to rapid development of thought, to appreciate the gradual development of ideas in prehistory. From the parables of Jesus to the capability for total self-destruction has been a mere two thousand years: from the act of standing on the hind legs to the birth of Christianity took four million years - two thousand times two thousand years. Within that timespan, a couple of million years passed before the invention of simple stone tools, and then possibly another half a million years until the domestication of fire.

The earliest hard evidence, from a cave in South Africa, tells us that our human ancestors were continually and habitually using fire one million years ago. That they did so in the same place over a long period means that they knew how to start a fire themselves, or keep fire going, rather than relying on the chance of a convenient local wildfire. The South African fire-places contained remains of bone that had been burnt, though there is no way of distinguishing between cooking and rubbish disposal in this context. But somewhere along the line of our predecessors, humans learnt to cook.

Fire technology had arrived, and would stay, becoming more sophisticated as the centuries rolled along. The cooking fire was a multi-purpose tool: it kept predators away; provided light and heat; transformed raw materials into delicious meals. It was the heart of every enlightened encampment. It was the campfire.

The campfire never went out. It was passed down the generations in every human family until all the peoples of the world had their own fire technology, their own ways of kindling it and utilizing it. More than twenty years ago, Ray Mears' bushcraft documentaries showed a white man taking primitive fire-making skills back to remote tribes who had abandoned laborious stick-rubbing in favour of using a cheap disposable lighter. The onlookers' expressions are generally of mild curiosity at first, but without exception their interest quickens with the first whiff of smoke, broadening into huge grins of delight as the tinder of grass or moss in Mears' cupped hands bursts into flame.

I believe the appeal of the campfire to be visceral: deep down inside practically everyone, with the probable exception of pyrophobics. Nowadays, for most people in the developed world, the experience of fire is sublimated: we have tamed it to the extent that few see it in its naked state, although we rely on concealed fire for a great deal of our energy requirements and everyday travel. The closest that some come to raw flame is lighting a candle, or cooking with gas. Yet even these common, everyday actions have a direct link to the primitive functions of the campfire: lighting a candle dispels the darkness, preparing a meal on a gas stove is just a sophisticated version of cooking on an open fire. The rise of barbecue culture is a sign of the subliminal human need for fire, and for the male of the species in particular, lighting the 'barbie' is an opportunity to demonstrate long-suppressed firemaking skills. Or otherwise.

Our more recent history is packed with examples of how important the campfire was to travellers in many situations. Classic within this narrative is the experience of those marooned on islands for years on end, unable to leave. Two instances from the eighteenth century will serve to illustrate the vital role of fire in such circumstances - one well-known story, another less so.

In 1704 the navigator Alexander Selkirk asked to be put ashore at a remote island in the southern Pacific Ocean, having had a dispute with the captain of his ship. The island was called Isla Más a Tierra, which means 'The Island Closer to Land', although more than 400 miles of empty sea separated it from the Chilean mainland. Once his ship, the Cinque Ports, had disappeared from view he began to learn how to survive in strange territory, completely alone, with just a few possessions.

At first he foraged along the shore, but before long set up camp some way inland. He built himself two shelters and hunted wild goats for food. Key to his survival, however, was the fire that he lit and kept burning night and day, taking care that it was not often accidentally extinguished. Naturally he used the fire for cooking, but one account indicates that at night he would burn pimento wood, which produced a candle-bright flame, allowing him to read his Bible.

Selkirk's story was later fictionalised by Daniel Defoe in his novel *Robinson Crusoe*, though it must be said that the campfire does not feature prominently in this narrative, only coming into significance some time after Crusoe's shipwreck, when he learns to make pots and bake bread. Defoe's book is rather odd and inconsistent, for all that it has been loved by past generations of children, and the campfire's want of importance should not be taken seriously.

For Selkirk himself, matters were quite different: the fire must be kept burning at all costs, for it was literally a lifeline. The first vessel to visit the island after his arrival proved, unfortunately, to be a Spanish pirate ship. He narrowly escaped capture, but the crew destroyed his shelters, and - an even greater act of inhumanity - put out his fire. Fortunately he had the means to start another one when the pirate ship departed. Selkirk survived, and was rescued by an English ship after four years of solitude.

The second story is even more remarkable, although it almost disappeared from memory. In November 1776 the French navy corvette La Dauphine dropped anchor near a tiny bare island, marked on the map as Ile de Sable (Sand Island), 280 miles east of Madagascar in the Indian Ocean. The ship's captain, the chevalier de Tromelin, sent a boat ashore, where its crew were astonished to find an encampment of seven women and an eight-month-old baby.

The women, dressed in clothes woven from feathers, turned out to be former slaves from Madagascar, the only survivors of a shipwreck fifteen years previously. The details of their story are patchy and contradictory, but it seems that they survived by eating turtles, seabirds, and shellfish, which they cooked over a fire kindled from their ship's salvaged timbers.

That ship, L'Utile, had left Bayonne in France in November 1760, bound for Ile de France - modern-day Mauritius. The captain, Commander La Fargue, stopped off at Madagascar to take on supplies, and he also decided to embark an illegal cargo of sixty Malgache slaves. The Utile was subsequently blown off course by a storm, and wrecked on the reef surrounding Sand Island. Almost all on board managed to make it ashore, though every one was injured in some way. They salvaged as much as they could of their ship, and set about making the means to survive in a hostile place.

They dug for water, finding it at the second attempt, and made a shelter for the fire that would cook their food. The French sailors, with the help of the slaves, then set about making a boat in which to escape the island. The task took more than two months, and when it was ready, every able person helped to launch the craft. However, not everyone was allowed to board it. The French sailors crammed themselves in, together with some supplies, and cast off, promising the slaves that they would send help when they reached civilisation.

One account claims that none of the sailors survived, another states that they did reach Madagascar, and went on to Mauritius, where they reported the shipwreck, but the governor of the island was incensed that La Fargue had taken slaves aboard, and refused to send a ship to rescue them.

Whichever version of events is the true one, it made no difference to the slaves, who waited in vain for two years before some of the men attempted an escape by raft. It is thought unlikely they survived. Ten more years went by before a passing ship noticed signs of life on Sand Island, and sent a small boat to investigate. The boat crashed on the reef, and one of its occupants swam back to the ship, while the other was marooned on the island. This sailor, together with the last three remaining male slaves, built yet another raft and left the island, never to be seen again.

According to their own account, as told to de Tromelin, the women who were left continued their solitary existence for about another three years until his arrival. The presence of an eight-month baby, however, suggests that at least one male must have been left behind, though no trace of him was found. Details are elusive, for although the chevalier eventually made his report (and mention of it is made in the French naval archives), the document itself has not yet been found. Most of what we know has been passed along by word of mouth, and the names and identities of the survivors have almost gone into oblivion. Sand Island was renamed Tromelin: that much has endured intact to the present day.

The one thing that seems certain in this remarkable story is that the survivors of the shipwreck lit a fire, and kept it going for fifteen years. It was so important that they built a shelter for it, to keep the flame alive through all kinds of weather. We can only imagine the hardships and the bitter despair of the slaves, ironically freed from captivity whilst trapped on their island, yet it is a sure guess that keeping the campfire lit was not only a practical part of everyday life, but also a means of comfort, and a glimmer of hope in circumstances where hope had no right to exist.

I put a match to the paper and kindling in the dog grate on our first night staying in the cottage. Although we'd cheated a bit by having a few celebratory fires in the time before the hearth was in place - before we even had such things as doors and windows - I had wanted to keep the first proper firelighting for our first evening in our new home.

I'd like to be able to say it was a wonderful success, but it wasn't. The flue was damp and took a lot of heating up, so smoke billowed into the freshly-painted room. Karen gritted her teeth and tried to smile as I struggled to get the fire to produce flame rather than billowing smoke. Eventually it did get going, but I knew I had a lot to learn.

I discovered that the dog grate was too far forward on the hearth, and pushed it right to the back, into the curve of masonry that the original builders had made. There is no throat at the base of the flue, so there is a tendency for smoke to billow in the space behind the lintel as the random particles encounter the narrowing-down of the flue itself. I re-learnt the importance of starting with a fast-burning kindle, gradually adding small pieces of dry tinder to minimise smoke and burn hot, and only putting large logs on when there was a glowing base to the fire. I learnt to keep the door between the kitchen and the living-room open during the initial kindling of the fire, to help the draught. And ultimately I learnt that the whole thing - hearth, flue, fire - is something like a living creature, with its own ways and needs and moods. Smoke rarely escapes into the room now.

At the end of all the learning I can now enjoy my role of firemaker, firekeeper. I am a servant of the hearth, providing all manner of fuel: kindling, faggots, sticks, logs. I rake out the ashes and sweep the hearth. So far I haven't swept the chimney, and wonder if I ever will need to, as we burn only very dry wood. Each evening we settle down in front of the fire, and whatever else we are doing, at some point we find ourselves staring into it.

Living flame is hypnotic: we are mesmerised by the play of colours, from white through yellow to blue, and the sounds of wood being consumed - hisses and crackles and soft shrieks. We cannot help being drawn to the fire, and in so doing we travel down a long road into our subconscious past, making connections with thousands of generations of ancestors who also stared into the flames, and went inside themselves.

Eighteen: Touch Wood.

I came across the word *touchstone* in an article I was reading, and realised I didn't know what it meant. I understood the metaphor, a touchstone being the test of the genuineness or value of something, but there had to be an original for the metaphor to spring off. What I hadn't known was that a touchstone was a piece of fine-grained stone which could be used to assay the purity of gold and silver. A piece of metal was rubbed across the stone, leaving a coloured mark, which could then be compared with the marks of test samples of known purity. Touchstones were in use more than five thousand years ago.

I looked up the definition in several reference books and dictionaries. In the OED the entry after touchstone is *touchwood*.

This is the kind of serendipity that comes only from a book. Ask your computer for the definition of a word and it will give you the meaning of that word and nothing more. Looking up a definition in a dictionary is an altogether more hazardous enterprise. As you flip the pages you can't help noticing other unrelated words, which may be odd-looking, or comical, and your curiosity leads you learn their meanings. It can take a while to get to the word you set out to look for.

I was struck by the proximity of the two words, rather than their true meaning. (Touchwood is dry wood that can be used as tinder to light a fire.) I took the terms metaphorically, and their conjunction spoke to me of my own working life: a transition from touching stone to touching wood.

For almost forty years I spent most of my work time handling stone: rolling it, hefting it into place, striking it with a hammer, sawing it, chiselling it, cursing it, loving it. During most of that time if I touched wood (apart from making firewood) I would end up botching any attempt to join two pieces together. I was quite good at stonework: I was rubbish at carpentry.

Faced with the prospect of joining many pieces of wood together as we rebuilt the cottage, I knew that my skill level had to rise from pathetic to at least adequate, and so it proved. Having good woodworking tools was a key, but I found that if I took my time, and measured twice, cut once, the results were acceptable - sometimes even rather good. And there was always the white miracle of decorators' caulk to mask any slight transgressions.

I am at the stage where I still have stonework to do, though much less than previously, and I have much more work with wood in store, and see this only increasing as time goes by. I have fallen in love with wood, yet not out of love with stone. I am keeping two equal but opposite passions alive in my heart. Actually they are not equal: I have a working lifetime of experience with stone, but only the beginnings of discovery of my friendship with wood.

In following this train of thought I let a finger of light into that dry tinder, since I have in recent years found myself using the saying *touch wood*. This may well be part of my experience of and awareness of wood, but it still surprises me, because I claim that I am not superstitious.

It is probably the most commonly-used saying to ward off bad luck, or rather to entice good luck. "The weather looks fine for the parade on Saturday, *touch wood*..." the speaker will say, matching the phrase with the action of tapping their knuckles on some wooden object, or in the absence of anything suitable, the side of their own head. Persons over fifty are most likely to utter the phrase: in country areas I have heard younger people use it as well.

Even a cursory search reveals that nobody can definitively point to the true origins of this practice, and there are a multitude of conflicting opinions on the subject. There is good reason to believe that the habit of touching wood, in the sense of the trunks of trees, goes back a long way, but it seems that the phrase itself is not at all ancient, possibly being as recent as 1899, according to the OED.

There are suggestions that it all has to do with the spirits that inhabit the material world, and that whilst uttering some statement of intent you should tap on the nearest tree so that the spirits can't hear what you are boasting about. On the other hand, some suggest that knocking on a tree wakens the spirits and invites them to support your intentions. A folk practice as adaptable as this must be useful in an emergency.

Our forebears lived among trees in a way we can only now imagine, and they invested the native trees around them with sacred significance. Oak, ash, hazel, and hawthorn were all treated with respect for their protective powers, as folklore amply demonstrates. Hawthorn, for example, carries a widespread and deep-seated 'health warning' about the consequences of bringing its flowers indoors. Nowadays most of the folklore is forgotten, but a residue remains, of which 'touch wood' is a clear example.

I say that I am not superstitious, but in thinking about this I have had to confront an uncomfortable fact: for the better part of my adult life I have believed in omens. The most consistent omen, for me, has been the sight of a bird of prey around the time of some major decision or event. Ever since I became interested in birds, the peregrine falcon has been a favourite, so it is hardly surprising that the sight of one should in itself be an event of some significance. And because seeing a falcon is notable, it heightens awareness of life, so that anything important that happens around the time of the sighting becomes inextricably linked with it. At least I think that's the way it works, but maybe there's more to it than that.

Perhaps the most striking example of the falcon as an omen was one May, when Karen and I holidayed in Kintyre on the west coast of Scotland, and on this particular day visited the island of Gigha. In the afternoon we walked across difficult terrain to the west side, and turned north, along the top of the cliffs. Before long the distinctive call of a peregrine captured all my attention, and soon there was a pair, circling us at a distance, calling incessantly. It was obvious there would be a nest somewhere on the cliffs, and to save disturbing them too much we continued on our way.

But almost straight away we came across the nest - in fact it would have been hard to miss it, as it was right at the cliff-top, where a gulley cut into the rock-face, and a rib of rock provided shelter from the westerlies. There on a ledge was a large messy nest, packed to bursting with three white downy chicks. Five peregrines in one place - a record sort of omen, surely.

A couple of hours later, as I was driving us back to our holiday let, and turning right off the main road, a tipper truck attempted to overtake, and ploughed along the offside of our car, before careering off the road, with us following in its wake. Apart from subsequent whiplash-type effects, neither Karen nor I were injured, even though the car was badly damaged.

I knew however that if the impact had happened just a second or so later, when we would have been squarer on to the truck, we would most likely have died right there, or at least have been seriously hurt.

The peregrines and the car-crash became linked that evening as I thought about what had happened to us. I had been so excited at the encounter with the falcons that I'm sure it didn't cross my mind it might be a portent of the future. In fact that is probably the case with every such sighting: only afterwards are the raptor and the event conjoined in memory. And if I had thought that seeing the falcons was an omen, what would I have done about it? I would still have driven down that road, at the same time, for the simple reason there was no alternative route back to the holiday cottage. There was nothing that could or should have been done. Birds of prey do not suggest what sort of event heads my way, good or bad, nor when it might happen. So what exactly do I mean by granting these occurrences the status of omens?

At a simple level, the conjunction of events that give rise to an 'omen' are pure coincidence, and the mind that does not wish to explore further can school itself to think accordingly. But my lengthening experience of life has shown me there are coincidental events that give rise to moments of recognition or correspondence. Such moments have ranged from a low-level awareness of a pattern, of events repeating themselves or forming some kind of progression, to a probably whimsical feeling of oneness with the universe. In between there are more realistic moments of recognising another person's experience as identical to my own, even to the extent of imagining another's life and thinking it to be parallel with mine, corresponding on more than one level.

I imagine that I may have begun to say *touch wood* around the time I first read Roger Deakin's *Wildwood*. I felt a great affinity with him, and knew from the start that this was going to be a very important book for me and that my own experience of trees and woods would in some way spool out in a continuation of his work. Trees became more and more the focus of my attention, and I, through them, became increasingly rooted in a natural and sustainable order of things. I have no wish to be mystical about this (it is an essentially pragmatic experience), but there is no doubt that an awareness of the vitality and durability of trees gives rise to a respect that will express itself sometimes through touch. And if the utterance of a phrase and the tapping of knuckles on wood gives a sense of connection with not only trees themselves but also the countless generations who have shown them respect, then that can only be a good thing.

I have never sought for omens, in the sense of the ancient Greeks' practice of reading the entrails of sheep, or even the Roman augur's attempt to interpret the will of the gods by studying the flight of birds. I would never consider entering a fortune-teller's salon to have my palm read.

My omens do not warn of doom or tantalise with the prospect of good fortune. They do not instruct me as to what path to take. They come out of a heightened awareness of the world around; they mark the progress of life lived with eyes wide open. I recall the powerful emotion of recognition I experienced, reading the beginning of Sara Maitland's *A Book of Silence*, where she recounts her experience of hen harriers. The first time she went to look at the ruined cottage that was to become her home, a pair of harriers were sitting on the garden wall. And then from time to time she would see one passing close by - 'pure gift', she called it.

For me, omens derive from such moments of synchronicity, of unrelated events crossing paths. I find myself nowadays looking upon them purely as encouragements, as affirmations. The course I am embarked on may be for good or ill, and I cannot know the outcome of it. Nevertheless I am embarked on it, and the glimpse of a peregrine or a barn owl serves only to say *carry on: do your best*.

When our small convoy of luton van and pickup truck arrived at Paradise Cottage at dusk with the last load of our belongings, it signalled the end of a difficult and exhausting move. As I climbed out of the truck a sparrowhawk flew out of the field opposite and into a tree almost above my head. It quickly spotted me and raced off through the wood, but I had not missed seeing it.

On the day we came to look at White Croft Cottages, a buzzard was perched in the top of the solitary old rowan halfway along the March dyke, surveying its domain. It kept an eye on us as we explored the buildings, and watched more intently as we made our way across the field. When we got too close it slipped out of the tree and glided away along the bottom of Clonyard Wood.

Nineteen: Coppicing.

Spring Wood, early March. Before I start the chainsaw I can see and hear that the small pools caused by dolloping are alive: the water boils with activity, there are splashings and ploppings everywhere, and the underlying deep churring of spawning frogs.

They first spawned on a mild day in mid-February, the sort of day that is too early for its own good, that makes chaffinches think of nesting. Five days later came snow from the east, and searing cold that put paid to the frogs' first attempt. But now the snow has melted back, just a few dirty-looking drifts in shady places, the temperature is rising steadily, and it is spawning time once again.

As I work the coppice I am conscious of being a disturber - hopefully not a destroyer. The frogs writhe on top of rafts of spawn, which in some cases almost fill a pool entirely. Their glossy skin colour is indeterminate: sometimes it looks very dark brown, then it might be grey-blue, according to the light. Underneath they are light grey, mottled with darker spots. It seems we may have different frog-families here, as the ones I encounter on Orchard Bank during hay-making are yellow-green or yellow-brown.

As I work closer to each pool its inhabitants sink their bodies below the surface, just the top of their heads and their intense eyes showing - eight pairs in one pool alone. To them I am a silhouette against the bright afternoon sky, a howling death-creature coming ever nearer to their birthing-place. Perhaps to them I am their nemesis, the giant Grey Heron from Hell, with my long screeching beak that rips through wood, tossing trees aside, scattering bright chips across the soft peaty ground. The frogs are motionless, unblinking eyes staring up at my shape as I loom over them.

Although I can't hear above the chainsaw's whine, they resume their orgy of procreation as soon as I move away. Only when I switch off the machine do I hear the splashings and the churring: the urgency of spawning is too great to be delayed. It is the imperative of Spring yet again: the very optimism of life itself. Before long early-seeding plants will be producing their superfluity, broadcasting willy-nilly into the wind so that perhaps a handful of seeds may find a lodging-place. Here in the pools the frogs are seeding water, splurging out abundant chances of life.

This is the first significant coppicing I have carried out. Last year I cut a small area, to check that it wasn't too early to begin the coppice-cycle among young trees, and was heartened by the results. It was clear that ground-flora would be vigorous from the start, so I deliberately cut the trees at 'stocking-top' height, to give the regrowth a chance not to be swamped. I knew that speedy-gonzales willow would be all right, but I was pleased to see that ash and alder got away, and the single trunks that I had cut were regrowing as four, five, six stems - exactly what I wanted to achieve. Roe deer had nibbled a couple of regrowths along the edge of the cut, but nothing more.

So last autumn I made plans to cut a much larger area at the top end of Spring Wood, where alder, willow, and some ash were all well-grown. These are the trees planted seven years previously, and most had reached about fifteen feet in height, with varied girths of trunk - alder being the stoutest, followed by willow, then ash. There are also some birch and hawthorn, and hazel along the margin, but I intended to leave these uncut, plus any other trees that have not grown so well, to give them another chance to flourish.

This is primary coppicing - the first time the trees have been felled, and for the most part they are single-stemmed. I had felled the cut over a day and a half, dropping all the trees in the same direction, so that they overlay one another. Now I begin to work up the coppice, starting at the end where I finished felling. This is a much more time-consuming task, added to which I want to clear out any plastic remnants - tree shelters, vole guards, mulch mats and the like - while I have easy access to them.

I cut the main trunk of each tree into five- to six-foot lengths with the chainsaw, working across the width of the cut. Then, with the billhook, I trim off side-branches, letting these fall around the exposed stools, the idea being to create a carpet of brash that will deter deer from browsing the regrowth. To begin with everything is a bit of a tangle, but soon a clear work-space develops, and the rhythm of the work becomes easier.

The trimmed poles are taken to either flank of the cut - over the fence into Top Field on the west side, and onto the raised path of Pipetrack Avenue on the east. Then the top branches are trimmed for any useable thin poles, and anything too small to be used is thrown behind to add to the carpet of brash. This process exposes the next layer of cut trees, the chainsaw is fired up, and the cycle begins again. Eventually the cut, which began with the felled trees all pointing north, ends up with interwoven brash pointing south, and the coppice stools just sticking out above the protective branches.

It is a much neater result than might have been expected from such twisty material, and I experience the glow of satisfaction that woodlanders must have felt for generations back, of hard work completed and a job well done. The other satisfaction is in the amount of material that has been produced. As the poles are carted up to the woodshed and stacked in a cord, my pleasure at the outcome increases with each power-barrow load.

This year's production is entirely for fuel. I can see that amongst it there are poles that could be used for hurdle-making or other woodland products, but that will have to wait for a future time of greater plenty. Everything that comes out of Spring Wood, together with the scrap willow that is unsuitable for basketmakers to use, will be burnt in the boiler that heats our home, or in the dog-grate fire that toasts our toes in the evenings.

The poles will remain in the stack until midsummer, drying out gradually in the wind and the sun. From July onwards a wet day job will be to work in the woodshed, processing wood into fuel. I'll begin with the scrap willow, which will end up as fire-starting material: sticks, kindling, faggots. There is a subtle difference between sticks and kindling. My way of lighting a fire is to begin with a faggot - a tied bundle of twigs - placed sideways across the middle of the grate, to help keep the infant fire as far back as possible. Behind the faggot I put scrunched-up newspaper, with some thin kindling criss-crossed over it. I set light to the paper, and as the kindling catches, I add sticks, which are slightly thicker and longer. It is a progression through ever-larger pieces of wood that is the hallmark of a successful fire.

Each scrap willow rod will typically provide two or three sticks, three or four pieces of kindling, and a contribution towards a faggot. The thinnest bits (the tip ends of the rods) I push into a metal tube about a foot long and four inches in diameter and sealed at the bottom, until it is tight-packed. Then I pull the bundle upwards until its mid-point is above the rim of the tube, and wrap a wire potato-bag tie around the bundle. The ties have a loop at either end, and a special tool is inserted through the loops and pulled outwards, tightening the wire tie and condensing the bundle still further. Faggots burn hot and fast: they were once used to heat bread ovens, quickly producing the required temperature, then the ashes were raked out and in went the bread.

When I have processed and bagged up the kindling, sticks, and faggots, my attention will turn to the coppice poles. The majority will be cut into lengths to fit the firebox of the boiler. Only the thickest poles will be sawn up individually, most of the thinner ones I will stack in a wooden cradle I have designed for a purpose. It has two pairs of uprights that are my chainsaw bar's length apart, and between them I can fit perhaps twenty or thirty poles, sticking out of the cradle at either end. The poles are held firmly in place by their own weight, and I can saw down through all of them at the points corresponding to the dimensions of the boiler's firebox. It takes longer to stack the cradle than it does to saw the poles: in just a couple of minutes I end up with a slew of logs cut to length.

The logs will then be stored along one end of the woodshed, between two large pallets fixed upright at right angles to the wall of the shed. There they will continue seasoning until they are dry enough to be called firewood. With small-diameter logs this can take as little as six months under well-ventilated cover, though it is more usual to take a year or more to season properly. It is quite likely that before the first of these new logs catches fire I will have coppiced another area in our woods, the regrowth on the stools in the previous cut will be three or four or even five foot high, a new generation of frogs will be spawning in the pools, and so the beautiful cycle goes on.

Twenty: Garrons.

The black ponies graze Top Field side-by side, their winter coats dappled dark and silver in afternoon sunlight. They step forward in harmony, necks arched down at the same angle, teeth ripping at the close sward. As they pass in front, when they are exactly side-on to me, they merge into one.

Their body shapes are almost identical, but not quite: there are features that distinguish one from another, allowing us to identify them from the other side of the field. Finn, the elder, has a heavier jaw: his half-brother Mister has dreadlocks across his forehead. They are from the same bloodstock, their sire Underwoods Drummer, but out of different mares, mother and daughter. Finn is out of the daughter, Mister out of the mother, which means - in the strange world of horsebreeding, and expressed in human terms - the younger is uncle to the older.

'Friendly' would not be a word to best describe their dealings with us: 'aloof' would serve better, but their interest in us is kindled when food is on offer. They are fed a bowl of mash every afternoon, at five o'clock in the summer, four in the winter. We joke that they must have a clock somewhere, because without fail they are waiting at the feeding-place just before time. After their feed they get access to a fresh patch of lush grass, as their pasture is doled out by strip-grazing, moving a three-strand electric fence across the field yard by yard, twice a day.

They are Fell ponies, and bred deep into their bones is the ability to survive and even thrive on poor ground. We say they would grow fat on thin air. At Garronfield the grazing is hardly top-notch, but it is far better than the garrons need, and left to themselves they would gorge until bloated, then suffer the consequences, the main one being laminitis.

In their second year here, one morning in early summer Karen woke to hear Finn yickering. He was standing with his front legs well forward, trying to keep weight off the hooves that were inflamed and causing him agony. That was the start of an annual routine of controlled grazing, the making of a grass-free paddock for quarantine use, and a heightened awareness of grass-growth and weather - sun after frost, for instance, produces high levels of sugars in young grass.

Blood-brothers they may be, but by temperament they are quite different. Finn is calm and steady, almost never jumpy, and he scrutinises the world around him, and especially our baffling activities, with kind-seeming brown eyes. If he had been brought up to it from a young age, he would have been superb working with disabled and disadvantaged children. When my young family visit, and want to see the ponies, it is clear that he puts back his shoulders (as it were), and becomes extra vigilant as the noisy little people dart about, despite our urging them to be calm and quiet. Unlike them he makes no sudden movements, and submits to their handling of him, and is not startled when they squeal as he gently takes a piece of carrot or apple from an outstretched palm.

Mister is much more wired. Karen has a piece of metal in her arm, souvenir of his early unpredictability when she was first riding him. As she leaned to turn him in the field he bolted forward at full speed, and they soon parted company. The radius in her left forearm was broken, and need ironmongery to get properly fixed.

There is no malice in him, though when younger he did tend to use his teeth to sample the world around him. He has matured now, and the unpredictable has mutated into stubbornness. He is lazy; he does not want to start work; he will refuse to go forward. But once he gets going he is good, and tries quite hard, and Karen gets much satisfaction from working with him.

When we took them on we were inexperienced in dealing with horses, and quite naïve about how they would feature in what we imagined of the future. We didn't have so much as a blade of grass to feed them, and asked Jane Barker to keep them for us at Heltondale until we had found somewhere. Finn was four years old, and the idea was that he would be the one for Karen to ride, while Mister, a foal, would be brought on to become the horsepower alongside me in the dream wood of my own that I so longed for.

In the end our plans were turned upside-down. Laminitis incapacitated Finn just around the time when Mister was being backed, and so the younger one became the riding horse. Finn was, from time to time, able to be ridden, but generally he settled into a quiet life, grazing contentedly, sunbathing, contemplating the world around him.

Their pasture was surrounded in the early days by infant woods, with no need for a pony to drag coppice poles to the stack. And when the work did start, lameness was still a problem for Finn, so I used our tracked power-barrow to haul whatever small amount of timber there was to be had. The machine had a light tread on our boggy ground, and was highly manoeuvrable in tight spaces. With a tinge of regret I came to realise that horse-logging was an unlikely prospect, at least for the time being.

In his 1955 memoir *Coming Down the Seine*, Robert Gibbings describes a visit to the caves at Lascaux, in the Dordogne, which were occupied in prehistory by cave artists, who covered the limestone walls with paintings of aurochs, deer, cattle, and bison, as well as big cats, a bear, and a rhinoceros. The cave-system consists of several galleries: Gibbings notes that in one smaller gallery there were images of stocky little ponies in various colours. He made a drawing to illustrate his book: in it there are horned cattle and deer, and a pair of stocky black ponies.

You could not visit the caves today. They were discovered (in modern times) in 1940, opened to the public in 1948, but by the late 1950s signs of deterioration were seen, and they were closed in 1963. The caves at Pech Merle, east of Cahors, are partially open to the public still, and their paintings depict woolly mammoths, cattle, reindeer, and both single-colour and dappled horses. The artworks are reckoned to be 25,000 years old.

Further east, in the Ardeche, a more recently-discovered cave contains the oldest paintings. Chauvet was first explored in 1994, and subsequent examination indicates two periods of occupation: around 37,000 to 33,500 years ago, and then 31,000 to 28,000 years ago. On the walls are depictions of the usual herbivores, including horses, but also predators - lions, leopards, bears, hyenas. A carbon dating in 2020 indicates that the oldest painting was created 36,000 years ago.

There is a particular significance to the presence of spotted or dappled horses in the cave paintings. It seems that a wider variety of colourings is an indication of domestication, and that prior to this horses were either black or bay (brown with a black mane and tail).

I watch Finn and Mister grazing side-by-side, necks arching gracefully, and glance at Robert Gibbings' woodcut illustration, seeing the same arch, captured perfectly by a 'primitive' artist, so many millenniums ago. A shiver passes through me as I recognise a thread that runs from the Upper Palaeolithic to the Anthropocene, through the ages of creative humankind, that can be faintly seen, in certain lights, as a hint of bay in the coats of our stocky black ponies.

Twenty-one: The Mortality of Pheasants.

He steps nonchalantly towards the scatter of seed under the bird-feeders, as though he has never been away. He is a cock pheasant, immaculately plumaged in an iridescent palette from emerald to rose pink, with a bold white neck collar and blood-red wattles. If I have got the right bird, the last time I saw him was nearly eighteen months ago.

Pheasants have been visiting our food offerings since our first spring in the caravan. Then it was a single male that I nicknamed Limpy, because of his curious gait. He visited for five consecutive weeks, then disappeared. I found his body across by the burn: he seemed to have died of natural causes, as there was no visible damage.

The following March another male appeared, to be joined by a female a couple of weeks later. This hen was a particularly pretty specimen, sleek and lithe, with very fine mottling all over her plumage. Over the next few weeks this pair were joined by another couple, and for a while we hosted the four birds every day. Then through August it was down to a single pair: the pretty female and her partner had disappeared.

But she hadn't gone too far away. In early September I was walking along the fence-line beside Paradise Meadow, which was now knee-deep in grass that would be the ponies' winter grazing. I was alarmed by a sudden venomous hissing coming from the field, and turned to find its source. It was the pretty hen, darting towards me with wings spread, feathers ruffled, and beak agape, emitting a startlingly loud warning.

Straight away I saw the reason: behind her the grass was alive with chicks, mottled fluff-balls squirming into hiding. There were too many to count. I thought at the time that it was late in the year for such a young brood.

A few days later I got the chance to count the chicks, when the hen brought them across to the feeding-station. They approached up the track alongside Spring Wood, the hen pacing slowly along the middle of the track, while the chicks leapfrogged from one patch of cover to another, for all the world like a band of soldiers creeping up on an enemy position.

And for them I was their enemy, though I had nothing but friendly intentions. It was midday, and by the time the brood had reached the feeders there was nothing left: the sparrows had cleared up any spilt seed. I went into the outhouse, unseen by the pheasants, and filled a small tub with seed. When I appeared round the end of the outhouse, the hen gave a sharp call, and the chicks quickly vanished from sight into long grass. She, however, stood her ground, and mopped up the seed I broadcast for her. From inside the caravan I watched as the chicks stole out of hiding, one by one. There were eight of them.

The following week the brood were joined by the male, and then another adult female, so that I became accustomed to the sight of eleven pheasants at close quarters through the living-room window. But late in September the weather changed. At the start of the month's final week there were seven young, by the end of that week only five. Heavy rain and falling temperatures were taking their toll, and by October 8th there was just one youngster left, who survived for a few more days before succumbing to the weather.

It is becoming more acceptable to admit that humans are not the only creature with emotions, and there are well-documented examples of apparent grief in a number of species. But if the pretty hen was grieving for her young, it was tucked away behind her speckled breast-feathers. She continued to visit the feeders almost daily, along with two males and another paler female, throughout that winter and well into the following spring.

The males were quite distinct from one another. The larger of the two was a handsome specimen with fine plumage. The other one, that I unkindly dubbed 'the runty male', was rather nondescript, without prominent wattles or a white collar, and dowdy plumage. Nevertheless, he seemed to be the pretty hen's partner, and could stand up to the other male, which happened more and more as that spring progressed.

In mid-May the pair disappeared, and nothing more was seen of them until July 4th, when the hen brought four chicks to the feeder. Although this was a far more promising scenario for successful rearing, none of the chicks survived to the end of the month. They might well have been eaten by badgers or foxes, but I had a lurking suspicion that bad parenting might have played a part. One morning I found that the hen had arrived at the feeders with three chicks, yet I could hear the beeping of a fourth somewhere. I searched around, and found the missing chick on the far side of the caravan, out of sight of the others, but clearly audible. The hen seemed to pay no attention to the youngster's calls, concerned only to mop up the fallen seed as usual. It was left to me to drive the chick round the end of the caravan, to be reunited with its careless mother.

The adult pair stayed with us throughout the winter and most of the spring, then, just as had happened previously, they disappeared in late May. On July 10th the hen reappeared, this time with seven chicks. She lost one a few days later, and I had the feeling that events were going to repeat themselves, but I was wrong. This time the chicks managed to stay alive right through to October. Out of the six, there was one male, and most of the females were very pale in colour, unlike their mother.

One morning a loud *Cock-up! Cock-up!* came from the direction of the feeders. I looked out of the caravan window to see the scraggy looking male youngster underneath the peanuts. I imagined a startled expression, an 'I-didn't-know-I-could-do-that' look. He had found his voice, and that was the last I saw of him for several days.

When he returned he had completed his moult and was now graced with an approximate version of male pheasant plumage, which over the next week attained its full splendour as he came of age. He was quite different from his supposed father, the runty male, being in appearance more like the other male that visited us from time to time, and I couldn't help wondering if there had been a bit of infidelity earlier in the year.

Then he disappeared yet again, and did not return. He may well have moved on, or rather *been* moved on, for the crime of being more handsome than the runty male. He was the first of the youngsters to leave: the females followed suit early in November, and it was back to the two adults, who were rejoined by one of the pale daughters mid-January. By mid-May there were just the two females, then they went. I expected a brood to appear in July, but it didn't happen. There was no sign of pheasants at all at Garronfield.

Early in the new year I mentioned the pheasants' absence to Hugh McTaggart of Roundfell Farm, and he was adamant that badgers were to blame: their numbers had increased strongly in recent years. Also, nearby Southwick Estate had ceased rearing pheasants for game a couple of years previously, and our visitors may well have been the last of the escapees from there. Whatever the cause, the loss of our half-tame pheasants was a source of sadness to us.

But now there he is, bold as anything, just taking a few steps away when I come out with fresh seed, which makes me quite certain that he is the cock who came of age while we were still living in the caravan, and has returned after a mysterious absence to a differently-sited feed station, outside the kitchen windows of the cottage. He arrived on March 12th, and less than a week later he has been joined by a pale female, who is very likely to be one of his sisters. She is more timid than him, but it looks as though they are forming an incestuous bond, which seems to be not uncommon in the bird world. Inbreeding is limited by the scatter of young after fledging, finding their own territories and fresh bloodlines, and any genetic abnormalities that might arise from its occurrence would be swiftly dealt with by nature's uncompromising cull, natural selection.

Incestuous or not, their pairing has been fruitful. It is lunchtime on May 13th, and the two pheasants have been absent for about three weeks. But now here comes the female, high-stepping round the end of the old caravan, and in her wake an untidy gaggle of chicks. There are a lot of them, scarcely more than a day old. We watch through the kitchen windows as they make across to the feeders, the female purposeful, the chicks milling all over the place, the cock bird eventually emerging from cover at the rear. We try to count them, with difficulty, eventually arriving at the same number: thirteen. A good-sized brood for a young hen - perhaps too good-sized.

And so it proves. By the end of that week they are down to eleven, down to ten the next week, and then in the third week mortality really takes hold. In the space of four days the brood declines from eight, to six, to just four. The weather is not to blame: it has been fine, dry and warm throughout. Perhaps magpies are the culprits, as a pair of them have been very active nearby.

The losses have occurred despite the adults being the most attentive pheasant parents we have seen at Garronfield. After that initial foray by the whole brood, the hen left the chicks in the cover of long grass at the edge of the track, warning them with a sharp *tchip* to stay put, while she sprinted over to the feeders to fill up on spilt seed. As they grew, the chicks became more unruly, and burst from cover and scuttled across to join their mother. Woe betide any jackdaw that came near: the hen fiercely drove off anything of a threat, while the male provided a rearguard, shepherding the youngsters if they tended to stray.

It is early June, and a disappearing act has taken place. In the space of two or three days the hen pheasant has vanished, along with two of the three remaining chicks, who are now well-grown enough to speak back at jackdaws. The male then turns up with just one youngster, and the following day is alone, crowing throughout the day from one of his vantage-points near the house. He can also be encountered at several places around and about, even stepping purposefully towards our gate from up the track to the road.

We speculate that the hen might be laying again, it being early enough in the year for a second brood. Perhaps she will have better success, but in any event the mortality of pheasants is another example of the optimism of life itself: splurging out a large brood in order for just one or two to survive.

Twenty-two: King of Birds.

On the year's first truly spring-like day, a wren is singing from the top of the birch nearest the house. His song is usually delivered from the cover of a bramble-patch or scrubby tree, the singer invisible, but today he is in plain sight, proclaiming his territory.

The most notable thing about his song is that it is amazingly loud and far-reaching for such a tiny bird - only the goldcrest is smaller, of our native species. The song has semi-sweet notes, but always there is a hard piercing trill as a dominant part of it. As well as being loud, it is fast: it has been recorded as 56 notes in 5.2 seconds, which is too rapid for the human ear, but presumably coherent to another wren.

The wren features as one of the first birds on my species-list for Garronfield, because on our initial visit we were challenged by one, down at the bottom boundary. They are feisty creatures: it has been said that if wrens were the size of swans, with that attitude, then humans would walk in fear. Now, especially in the breeding season, wrens will always give me a good ticking-off if I stray too near to their hidden nest. And a ticking-off it is, literally, with a hard *churr* and a sharp tic-tic combined.

Everything about the wren seems prodigious. The males, in spring, make several nests, and invite their prospective mate to inspect them and select her favourite. This is not entirely wasted effort, because often enough when the first of two broods have fledged, the male will take the youngsters to roost in one of the other nests. When my pickup was in regular use, a wren constructed a large nest that consisted entirely of moss, on top of the rear axle, between Saturday teatime and Monday first thing. And now I think that one has designs on the woodshed, because he flies out of it whenever I come near. There are plenty of attractive nooks and crannies: unfortunately most of them are among the firewood, or my stock of willow.

It is suggested by ornithologists that a male wren will defend a territory of about two to three acres in extent, and certainly this seems to be borne out here. One summer Sunday afternoon walkabout found us being challenged vociferously at the bottom boundary, to the east of the Singing Place, where our only blackthorn tree had evidently become the hideout of a wren family, who were all intent on letting us know we were not welcome. Two hundred yards further on, in the middle of March Wood, we had a repeat performance from a second family, who were more spread out among the trees, and we inadvertently walked through the middle of the group, who, again, protested about our behaviour.

Perhaps loud-singing wrens drew too much attention to themselves in the past, for there was a long tradition in parts of Britain of 'Hunting the Wren', whereby youths and young men would beat the hedgerows, trying to kill a wren, which would then be paraded with much fanfare through the local community. This ritual would take place between the winter solstice and Twelfth Night, according to regional variations. Some folklorists think that the custom might date back to the Bronze Age, but in any event it was practised widely in Wales (on Twelfth Night), and Ireland as well as the Isle of Man on St Stephen's Day (Boxing Day). The Manx ritual is practised to this day, although thankfully the wren is now represented by a symbolic object.

It is beyond my comprehension why anyone would wish to murder such a neat and harmless bird. But there are stories from legend of the birds betraying one set of people to their enemies, by - guess what - making a noise. Perhaps revenge for that kind of betrayal lies at the heart of 'Hunting the Wren'.

And then there is jealousy. Folklore names the wren as King of Birds, the story being that all the birds once came together in a parliament to decide who would rule over all others. It was decided that whoever could fly the highest would be ruler. The eagle soared higher than any other bird, and just at the point where he could climb no further, the wren emerged from his feathers, where it had been hiding, and flew above him. The other birds were so enraged by this deceit they tried to kill the wren, to no avail. It is said that this explains why today the wren flies no more than a foot or two off the ground, and dives into cover as soon as possible.

Betrayal, revenge, deceit, jealousy - what a heady mix for such a small frame. It all serves to add to my appreciation of the bird, and I am always glad to see them close to us: especially glad after hard weather, which is difficult for them to survive. If any part of what we have built here can help in their survival, then for that, too, I am doubly glad.

And as it turns out, the woodshed did play its part. The wren that I had seen in there in early spring had not just been inquisitive, but industrious. Yet it was late July before I noticed what he had been up to. On the back wall of the shed hangs a coil of frayed old hemp rope, which has formed itself into a figure of eight. A stray piece of moss blowing in the wind caught my eye, and betrayed the tiny troglodyte's activities: in the top opening of the figure of eight was a compact nest of moss, with a round entrance the diameter of my middle finger. There was no sign of it being used: probably one of the many efforts spurned by the female as a suitable home.

Twenty-three: The Silence of Nature, the Nature of Silence.

At Garronfield there are times, most notably at dusk after birdsong has died away, when there is total silence all around. And later, on a cloudy night, the silence may be accompanied by utter darkness - if you turn your back on the dim lights in the new building up at New Farm.

The darkness may be more or less real, but the silence is a delusion. There is no silence in nature, there is only apparent silence. Usually there are sounds: the wind among trees, birdsong, the calling of fox and roebuck. The hare seems to be mute, but hunters say when one is trapped or injured it screams like a child - which has led some to give up hunting hares. On a normal day a parade of noises jostles around Garronfield. Few sounds are very loud: for the most part there is a harmony that is pleasing and soothing to the ear.

But at dusk, often enough, the apparent silence descends. The last blackbird gives over his fluting, the wind is still, no animals call out. Yet if we were privileged to host a barn owl here, and if it could speak to us, it would tell how, as it ghosts across our corners of left-long grass, there is not silence. Among the stems are the shrieks and scrabblings of voles and shrews, which the owl's asymmetrical ears are tuned to home in on, as it drops soundless out of the night on to its unsuspecting prey.

By day blackbirds quarter the bare ground at the front of the house, yet to become a garden. They hop along with head to one side, listening for the squeak of a worm in damp soil. A pounce, a darting beak, a tasty meal. There is no silence in nature.

If humans had the hearing that blackbirds do, I imagine that on an early spring day, as the land wakes up from a winter's sleep, we would hear the hiss of grass growing, invertebrates wriggling in topsoil, leaf-buds cracking apart, new leaves creaking open into daylight - an orchestra of sounds from the earth itself, and from all that is rooted in it.

In our evolutionary past we may well have had the faculties to hear these small sounds - together with keener eyesight and a wider sense of smell - but along the way we have lost the ability. What was no longer essential to survival atrophied, giving place to other strengths that put us on the path towards what we are now. There can be no hard evidence as to how well Neanderthals could see or hear or smell: the soft tissues of eyeball and eardrum and nostril do not survive, and there are no records to examine. But it's a fair guess that their senses were sharper than ours are today.

Silence, then, is not an absence of sound, but our inability to hear a range of sounds emanating from the earth, on this Earth at least. Silence is therefore a delusion, as I said previously, that we are content to live with. In other words it is an illusion, not in the visual sense, but that of a false notion. Those of us who love solitude and silence, and the well-being derived from them, are nonetheless happily deceiving ourselves: solitude is only relative on this populous planet, and there is no silence in nature.

But perhaps, perhaps. I think that for the last half-century I have believed in one moment of absolute silence in the wake of one particular event. I was fourteen when I entered my first poetry competition - a schools' competition organised by the Wordsworth Trust, to commemorate the bicentenary of William Wordsworth's birth. I wrote a sonnet about a peregrine falcon, and all I can recall of it was the last line: '...where a song-thrush paused over half-smashed snail'. The focus of the poem was on the silence that fell over the dale (rhymes with snail), after the falcon's stoop and successful kill.

My poem was highly commended, but failed to win a prize, because in the judge's opinion it was not quite in the spirit of Wordsworth. Any interest I might have had in the Lake Poet did not survive this judgement, and it was only many years later that I came to appreciate at least some of his writing.

What was undiminished, however, was my love of, and admiration for, the peregrine falcon - and that has continued to this day. I see a peregrine regularly here, and on a few occasions they have nested in the top of Bennel Wood, unnoticed until their shrill calls come to my attention, as they tempt their young into flight for the first time.

Twice I have been privileged to witness the aftermath of a peregrine kill. You often know when a sparrowhawk is out hunting when you hear the angry clamour of smaller birds in flocks, who harass the predator. By contrast, when a peregrine makes a kill, a profound silence comes into being. It is as though the earth has stopped turning for a moment; birds are stunned into stillness; nothing cries out; even the sheep on Barn Hill lift their heads from grazing.

I am no stranger to illusions. The poet Philip Larkin once said that he searched himself for them like a monkey searching for fleas, but I am content to let mine be, as long as they do no harm. So, awed as I am by the falcon's speed and power and deadly skill, something within me wants to believe that the earth *does* stop turning; that grass ceases to grow; that worms and beetles pause; that leaves hold still; and even my heart stops in that fierce convulsion, that utter silence.

Twenty-four: The Lonely Swallow.

I cannot help but feel a tug of sadness each time I see him, perched at the end of the gutter above the boiler-house door, a solitary swallow in the breeding season. All that way from Africa, and no mate at the end of it. If I wake in the early hours, pre-dawn, I hear him singing, from somewhere up on the bank, what I take to be an outpouring of longing. I wonder if the paired-off swallows in their cosy nests tremble and ache as they listen.

As I write this, I can hear the stern voice of a scientist telling me to get a grip, to cease from ascribing human emotions to much less complex creatures: the bird's singing is purely instinctive. But, I reply, if humans have evolved to possess a wide spectrum of emotional responses, why is it not possible for other animals to have developed their own range of emotions? A swallow's feelings may not be identical to human ones - perhaps not as complex, nor as intense - but nevertheless be appropriate reactions to life's circumstances.

In fact, we do ascribe emotions to other animals, in particular the big heavy-duty ones: fear, anger, and desire to procreate. Why not allow for a wider palette of more subtle responses: sorrow, joy, despair, pity? Actually, I believe that many people do just that, not out of ignorance, but out of the need to empathise. The ability to enter imaginatively into the experiences of others in order to understand them is a significant part of what makes us human, and the lack of it leaves us somehow less than human.

We know more about bird migration now than we did when it was thought that swallows overwintered in Britain by hibernating in the mud at the bottom of ponds. The British Trust for Ornithology has been tagging and satellite-tracking cuckoos since 2011, and this has revealed widely varying and previously unknown migration routes and destinations. The science may be improving, yet underneath is a deep layer of mystery: our understanding of the impulse and the trajectory is far from complete. Even now nobody knows exactly where house martins spend their winters. But for me the biggest mystery about migration is *why* - why expend so much energy and take such risks? Why did birds start migrating in the first place?

I have read that climate change, at the end of the last Ice Age, might have been an influence. But that can scarcely explain the arctic tern's annual circuit between polar regions, which almost seems like showing off - flight for flight's sake. Out of roughly 230 'British' birds - those that use these islands for one reason or another at some time of year - only about 120 species are resident, and even then some of these residents disperse from their breeding-grounds in winter. It seems there may be a deep avian reluctance to stay put.

It's having wings that does it, of course. Humans have from the earliest times envied birds their flight, and have sought to emulate it. But human flight has to be assisted, and even the neatest form is somehow clumsy in comparison to the simple spreading of feathered wings, the ability to take off at will from wherever. Mankind's efforts are but a poor imitation of birdflight, and a trace of envy must still reside within even those - paragliders and suchlike - who come closest to the pure natural freedom of birds.

Do birds enjoy flying? Flight is a necessity for almost all bird species: it has straightforward functions, such as finding food, keeping safe from predators, as well as for migration. But does the red kite, circling above me in summer thermals, on the lookout for food, nevertheless delight in its ability? It is impossible to know for sure, yet I would point to a handful of behaviours that might suggest such delight.

In upland places I have seen courting ravens plunging through storeys of air, spiralling within each other's wingspan, with the sound of swung swordblades, of flying arrows. It is part of a courtship ritual, yes, but they also do it after nesting is over. I cannot help but think they experience what we know as the elation of love, of faithful devotion, and express it through such spectacular tumbling through the air.

Again among the hills, after the breeding season, gangs of the sociable crows - rooks and jackdaws - also swoop and tumble in high flight, for no apparent purpose. They are not looking for food, nor heading to roost: they appear to be playing, and enjoying one another's company. Fanciful or not, it is what I think whenever I see such exuberance.

And then there is the flight and song of the skylark. Nothing speaks more of pure joy. The male is proclaiming his territory, of course, high above his mate on her nest - but does he really need to sing for sometimes twenty minutes at a time? It seems like overindulgence, as though having got the business part over - proclaiming *this is mine, this is mine, this is mine* - he is driven by another urge, just as deep, to revel in the delight of fluttering at the same point in the sky, outpouring the joyous cascade of his song.

The lonely swallow has given up, in the end. He has moved on, or died of grief, or been killed. The offspring of the paired swallows, their first brood, excitedly swoop around the buildings, exploring every opening. From time to time they fly into the house through the open side door, and become stranded in the passageway, trying to get out through one of the tall windows. I gently shoo them into a corner of the frame, where I can get my fingers around them without harm. They are not afraid to answer back with an angry chirrup, and even attempt to peck my skin. When I am outside, and open my hand, sometimes they sit for a moment or two before flying off.

Do they know, I wonder, somewhere deep inside themselves, that in a few weeks' time they will be undertaking a journey that will send them winging many thousands of miles to southern Africa? And that if they survive that migration, their overwintering, and the return journey, they will be back here next April, looking for a mate, to nest and raise their own young?

There is no appreciable weight to a swallow if you hold it in your hand. It is simply a marvel, almost miraculous, that such a scrap of life could take on a flight so epic and so dangerous. It is a mystery why they do it, and a wonder that they do. I shall miss them when they go, and will look out for them coming back to their birthplace.

Twenty-five: Shade.

In south-west Scotland there is rarely enough sunshine to make the seeking of shade an imperative, but the spring and summer of 2018 proved exceptional. For almost three months, from the beginning of May, we had dry weather and mostly ridge-to-ridge blue sky, from which the high sun beat down without interruption. There was a band of rain in mid-June, but it was the last days of July before the weather finally broke.

For the first time since we planted trees at Garronfield I began to appreciate and enjoy the shade they now provided. In full leaf, with canopies overarching the trackways, it became an intense delight to step out of the blinding heat into the cool greenness offered by our woods. Not everywhere was well-grown enough to provide shade, but one place in particular - in Spring Wood at the corner of Paradise Meadow, where three paths meet - ash and alder clasped one another overhead and stole the ferocity of the sun's rays.

In his poem *The Garden*, Andrew Marvell speaks of the human ambition to win laurels and live within their meagre shade, whereas in his beloved garden '...all flowers and all trees do close / To weave the garlands of repose'. Here the poet finds quietness, innocence, and solitude, and 'a green thought in a green shade'. Towards the end of 2018's heatwave, as temperatures in south-eastern England crept into the high thirties, the Met Office, in its paternalistic way (or should that be maternalistic?), took to advising everyone to seek shade between 11am and 3pm, but refrained from suggesting what sort of thoughts to have during that time. For my part, I had no trouble enjoying my own brand of green thoughts among the trees we had planted seven years earlier.

The noticeable increase in shade was as a result of strong tree growth throughout that spring and early summer. All species seemed to appreciate the dry weather, although by the start of July mature trees in the surrounding area were starting to look somewhat browned-off. Our coppice regrowth really burgeoned, ash and alder in particular putting on four or five feet of growth in a matter of weeks. The initial planting was mostly on the wet areas of our land, and it may well be that the extended dry spell conspired to produce perfect growing conditions, with sufficient moisture in the ground to keep everything green through the hot days.

Shade is an important component of woodland habitat, and the amount of it a governing factor in the type of ground flora. Oliver Rackham, in his book *Ancient Woodland*, suggests that continuous shade is not actually a feature of deciduous woods, and that most places within a wood will receive about a third of full daylight in spring. He also states that many plant species growing within woods do not actually require shade in order to flourish, but do so in a woodland setting because of the reduced dominance of light-demanding plants.

The most dramatic change in woodland flora takes place after a wood is felled or coppiced. In an earlier chapter I described the burgeoning of diversity following the dolloping process: the hoard of buried and suppressed seed having the sudden opportunity to burst into life. The act of coppicing has a very similar effect on the ground flora of woods.

Our plantings are still too new to have established much of a diverse ground flora. The early coppicings have allowed the more boisterous species - thistles, nettles, docks, umbellifers - to flourish and dominate in the initial abundance of light. But we know that other less-dominant species are present, though not highly visible, and we have reason to believe that as the canopy closes with new tree growth there will come a gradual change to a more varied and less invasive flora.

It is tempting to speculate on what plant species might gradually colonise our woods. We have planted bluebells, near to Holly Grove, and this species is the archetypal woodland floor plant. However, it is likely that it takes a long time for bluebells to establish an abundant presence, and it may well be that our intended coppice-rotation - seven years - is too short for that establishment to take place.

Primroses grow on the bank of the burn just downstream from March Wood, and it is quite possible that these plants may spread more widely over time. Ragged robin is present, and spreading slowly. Purple orchids appear in the less-shaded places, seemingly not out-competed by the vigorous growth of other species. Vetches, violets, anemones, celandines: all are there to be seen if you look closely along the path edges, and meadowsweet has become an increasing presence. The diversity is greatest in March Wood: Spring Wood is more vigorous throughout, and the undergrowth is coarser and thicker, but even here there are gradual signs of change in the ground flora, and an indication of how our woodland will develop as coppicing continues.

Twenty-six: Neglect.

I confess, with a hint of embarrassment, to being tidy. Tidiness is a trait that is often subject to mockery: if a cluttered desk is the sign of a cluttered mind, then what are we to think of an empty desk? (Albert Einstein said that). And the term 'tidy' can be used disparagingly, to mean petty, trivial, safe, uninspired.

The word itself has had an interesting evolution. It was originally the adjectival form of *tide*, the regular and predictable pulse of the sea. Tide then came to denote the orderly progression of seasons - Christmastide, springtide - and *tidy* developed to infer all that is methodical, well-arranged, neat. (There is a side dish of other meanings: a receptacle for odds and ends; a considerable amount, a *tidy sum*; in dialect usage, plump, comely, shapely).

We now understand that the orderly universe has disorderly chaos embedded in it. It is quite common to hear writers and artists speak of how they embrace chaos and disorder at the start of their creative process, then go on to shape and define the early inspiration. It is making a mess, then clearing it up; breaking things and reassembling them; making a din, then letting silence back in; splurging out a torrent of words, then picking a way through.

For many of those whose work is creative the end part of the process can be likened to a tidying-up. I know from my own writing that the final part is the least creative, but as necessary as anything that has preceded it. It involves a meticulous trawl through the slew of words, cutting out anything that is superfluous, misleading, or just plain ugly. But if the last part of the creative process is a tidying, this is not to say that the end result is necessarily neat and orderly - indeed, great works can often be *disorderly*, but not by accident, or through inattention or laziness.

The typical arc of creativity, from chaos towards order, is echoed in humankind's almost universal creative act: gardening. It begins with nature's prodigality, that optimism of life itself, meshed with the human optimism that sowing seed will produce a crop of food or beauty. At the outset there is a splurge, a spendthrift scattering of far more seed than is actually needed, which is nature's way of stocking the garden, echoed by the human gardener's thriftier scattering, which nevertheless usually produces greater abundance than can be used in the end. Then follows a process of thinning, of selection, of weeding out strangers, training, supporting, pruning, and so on. The aim is productivity, whether of beans or grains or fruit or flowers, and the underlying trait to achieve this aim is to be orderly, to be tidy.

As I have already described, in the early years of establishing the woods at Garronfield I was certainly orderly and tidy in my approach. But seven years on from that first planting I have discovered a rebellious streak at the heart of my nature. I find myself entertaining the concept of *neglect*, which feels to me like a deliciously wicked thought. Perhaps the idea has been germinating for a while, but it has burst into leaf - paradoxically - due to ash die-back. When the first signs of this appeared among our trees, we spent time cutting out affected branches and withered bunches of leaves, taking them out of the woods and burning them to destroy any spores. But by the second year of the disease being present, I realised that it was not feasible to remove all the affected parts, and resigned myself to letting things take their course.

This was at first a dispiriting time, particularly as it was clear that ash was one of the species really thriving here. But as the season wore on, and the ash trees did not die, merely dying back in some cases, whilst other nearby individuals thrived, I found myself embracing the mortality of trees, and in so doing gained a kind of release.

We know that many tree species commonly outlast the human lifespan several times over, and it is not unusual for individuals to have been well-grown before Jesus first wailed in his crib. The most ancient tree known to us at present is a Californian bristlecone pine, at 5070 years old, as of 2020. But in the young woods here at Garronfield a life and death struggle is taking place - a slow one, admittedly, but fierce in its own way.

The trees were planted at two-metre spacings, and still there is not much change within that pattern. Yet we know that in established coppice the stools are more likely to be around four or five metres apart, which means that a good number of the trees we see now are going to die, or at best fail to thrive. This process is already noticeable in its early stages in the first coppiced areas, where there is a marked difference in the height of regrowth at the end of the first season. Those stools that have put on less growth are most likely to be the first casualties of the struggle for life.

In essence, a natural thinning-out is beginning to take place. In commercial forests the process of thinning is a normal management operation, now largely carried out by sophisticated machines that have a low impact on the forest environment. My own low-tech outfit - chainsaw and billhook - is ever ready to carry out a similar operation in our new coppice, cutting out poor growth to allow the good growth to thrive even better, and working among trees in this way is something I love to do. But I am stifling the urge.

No: it is more, and better, than that. I am not stifling anything, but rather opening myself to a new way of seeing, which involves standing back and looking on as natural processes take their course. There is a synchronicity here with a passage I came across in Richard Jeffries' *The Toilers of the Field*, where he speaks of a favourite orchard near to his home: 'Pass when you may, this little orchard has always something, because it is left to itself - I had written *neglected*. I struck the word out, for this is not neglect, this is true attention, to leave it to itself, so that the young trees trail over the bushes and stay till the berries fall of their own over-ripeness, if perchance spared by the birds; so that the dead brown leaves lie and are not swept away unless the wind pleases; so that all things follow their own course and bent.'

There is currently a fashionable obsession with the idea of *wilding*, or *rewilding*. This is the concept of allowing swathes of countryside to return to a primitive, pre-agricultural state, combined with the re-introduction of mammal species that have become extinct in Britain. I am not an enthusiast. Rewilders often seem to wish human beings out of the landscape, but this is ignorant of the value of human life as part of nature, rather than separate from it.

I have no desire to live in a wilderness: I have chosen to live where there is space and freedom to be apart from my neighbours, but where there is the comfort of knowing that if I had trouble, help would be a few minutes away. And in my tame habitat there is wildness still.

There is a patch of it close to the kitchen windows, at the bottom corner of Orchard Bank, where a tangle of brambles and nettles surrounds an intertwined elder and hawthorn. The patch is dense as any jungle: the only way into it would be by wielding a billhook. The trees and the brambles are the cover the small birds dive into when the sparrowhawk appears. There may be nests in there: I wouldn't know. In March Wood and Spring Wood there are now such patches: brambles are taking over, it might be said, but I say they are taking their place - and their place is becoming *not mine*. I have the delicious sense that these dense tangled corners are out of bounds to me.

But that does not mean I ignore them: quite the opposite, for their very secrecy makes me want to peer in, to try and make out what is happening in there. This is the attentiveness that Jeffries speaks of: the neglect that allows nature free rein, yet carefully notes what takes place (or doesn't) in the wild places that are close to home. It is not the neglect of ignorance or indifference. It is, for want of a better phrase, alert neglect.

Twenty-seven: October.

Four magpies are perched in the topmost branches of an aspen, the tallest tree in Spring Wood. A stiff breeze from the northwest bends the treetop before it, and the birds are constantly in motion, opening and closing their wings, tails swivelling to keep balance. Piebald acrobats, sharp-eyed for something to investigate.

The breeze bows the triandra willows along the top edge of the wood, upending the foliage to reveal pale underskirts. A storm in mid-September ripped the leaves from the ash trees, but otherwise the woods are now well-grown enough to display an autumn palette of tints. Across the field, at the edge of March Wood, Karen's eye-catching planting - for spring blossom and autumn colour - is coming into its own, from the gold of field maple to the deep reds and purples of dogwood, and blood-bright fruits on guelder rose. Behind this striking fringe, the pollarded viminalis willows, which stand taller than the surrounding trees, shimmer like flames in late afternoon sunlight.

After a run of equinoctial gales come a few days of still mornings with heavy dew as the sun rises. On one of these mornings the river of gorse alongside the burn is covered in spiderwebs: gauzy napkins in terraces among the spikes, a filigree of lines connecting the layers, and all of it spangled with the dew. An early frosting, without the ice.

Mid-month, an ex-hurricane barrels towards us from the tropics. Daylight struggles to break through a veil of high cloud, coming up from the south, laden with Saharan dust. When the sun does appear, late morning, it is a deep dull red, and I can look at it directly for a few moments. It is calm, for the time being, just an occasional zephyr stirring leaves and grasses.

Around two o'clock in the afternoon there are the first stirrings of a warm wind, the dusty veil melts away and a strong sun shines in blue spaces between big clouds. Then comes a heavy downpour, lasting no more than a few minutes, and after it the sky is seamless blue, not a cloud in sight.

The first big gusts hit after four o'clock, and it becomes overcast, an ominous uniform grey. By dusk the wind is strengthening, with ever more frequent gusts, and rain splatters against the black windows. Karen is anxious, but we are hunkered down like a granite ship in the storm, wind growling in the chimney, shrieking in places under the eaves, our ventilation ducts burring and buzzing, and creakings from everywhere. The meteorologists have delighted in calling this event 'ex-Ophelia', making it sound like something that would strip a layer or two of skin off your face, as indeed it might if we were to step outside.

In the morning we find we have got through without damage, just a rearrangement of things not tied down, a dusting of desert sand over everything, smearing the south-facing windows, and fewer leaves on the trees than there had been yesterday. The storm is gyrating across the North Sea, heading for Norway.

In the wake of the storm, on an unusually still and summer-warm day, I am in March Wood, cutting back branches that obstruct the paths, when all at once I am surrounded by what seems like the tinkling of little bells. It is a party of long-tailed tits, quickly flittering through the trees, difficult to spot among the remaining foliage. It is impossible to say how many of them there are, but they seem to have picked up two or three blue tits along the way, or perhaps they are just straying through the blue tits' territory as I watch. In a few seconds they are past, heading towards Bennel Wood, and March Wood falls still once again.

That evening the bowl of a first-quarter moon, with Mars trailing behind, slips down a low trajectory towards New Farm, haloing for a while their newest building on the skyline before dropping out of sight. A little later the planet follows, along the celestial equator, below the dark horizon.

There are plans afoot to do away with British Summer Time, but I am quite fond of the clock-change on the last weekend of October. To divide the year into seasons, I follow the meteorologists' pattern, which is that the winter months are December, January, February, the spring months March, April, May, and so on. That these fixed seasons are not necessarily graced with the appropriate weather is just part of the idiosyncratic nature of the British climate. But if the year is to be divided in two, a summer season and a winter one, then the October clock-change has a particular appeal.

In March, the flip-forward of time seems like a call to action: to shrug off the sleeps of winter and confront the imperative of spring. By contrast the fall-back of time, that delicious lazy extra hour, is the signal to shift down a gear, for dusk will now come early, bringing to a close my outside work, ushering me indoors to study, write, and read. I can loosen my grip on the growing world, surrendering to the inexorable, as the green dividend withers and dies back into the earth from where it came.

And so, on the first day of the winter season, I take a walk around the place. The trees stand at all stages, from full foliage to bare. Some willows, as ever, cling on to their green leaves; oak and alder are yellow, red, brown; ash is stripped back to silver-grey branches. At the Singing Place a snipe startles up in front of me, and jinks away down the rushy bottom of Alan Carter's field. Roe-deer are moving below New Farm Wood, and the buck calls a warning, a hoarse bark that sounds like *Go! Go! Go!* But no, my friend, I am staying put.

Twenty-eight: This Land.

I am back at the Singing Place, to stand in the rain and observe The Silence on Armistice Day, one hundred years on from the end of the war that was supposed to end all wars. I try to think of all those that died in that war and all others since, but the millions upon millions of lost faces are too much to comprehend, so I concentrate on just one, my father's, whose own war began twenty-one years after the ending of the one we are remembering today.

My father did not speak of his wartime experiences. After he died I found, in the bottom of his wardrobe, the boxful of letters he had written to his mother throughout his time abroad. Most of them came from a series of prisoner-of-war camps in Italy and Germany, for he had been captured at Tobruk in Libya in June 1942. He was one of those who survived 'The March', the forced evacuation of POW camps as the Russian army approached from the east. He returned home in April 1945 with feet permanently damaged by frostbite. He died aged 78, and I have every reason to believe that what he had suffered during that war robbed him, in the end, of at least five years of life.

I have observed these two minutes of reflection for many years, always on my own, ever since I was moved and inspired to do so by farmer A G Street's memoir of agricultural life just before the Second World War: 'It is amazing the hold which Armistice Day still possesses. All over Britain this morning farm-workers set their watches by the church clock, and even where no warning siren could be heard, work stopped at the appointed time. Somehow I always think that for one man to observe the two minutes' silence when he is quite alone is almost more impressive than when a huge crowd does so. A ploughing team motionless on the hill-side, with the ploughman standing at attention between the plough handles, perhaps thinking of those dangerous days when he had little hope of returning to the land which bred him. A tractor which ceases its mechanised stutter at the behest of Armistice Day. A silent, motionless shepherd alone with his flock in the middle of a root field. An ancient woodman, standing motionless in a clearing with his billhook in his hand.'

Harry Patch had no truck with Remembrance Sunday - he dismissed it as 'show business'. He died in 2009 at the age of 111, one of the last survivors of the carnage of trench warfare. On the 22nd September 1917, he and three comrades were returning from action at Passchendaele when a shell burst amongst them, wounding Patch and killing the others, one of whom was his best pal. He observed a personal ritual on that date, remembering his comrades, for the rest of his life.

We have been watching again the documentaries made by the BBC in the late 1990s, featuring the reminiscences of First World War veterans, as the remaining few neared or passed their own century. There is a roll-call of names: Dick Barron, Robbie Burns, Jack Rogers, George Louth, Albert 'Smiler' Marshall, George Littlefair, Norman Collins, Arthur Wagstaff, Tommy Gay, Fred Francis. These were men who volunteered at the start of the war, who felt it was their patriotic duty, who were stirred by the sound of marching bands. Some of them were under 19, and lied about their age. After January 1916, when conscription was introduced, the roll-call continues, of men who were aware of the grim reality they faced, but were conscripted, or joined up because they felt compelled by public opinion: Bill Cotgrave, Alfred Henn, Fred Tayler, Archie Richards, Fred Hodges, Andrew Bowie, and Patch himself.

Their voices - all silenced now - still call out from the precious film archive. They record how they felt they had lost all sense of humanity, how the war was not glorious, but wicked. Strongest of all, perhaps, is Harry Patch's fierce indignation, calling out in his Somerset accent: 'In the end they settled it round a damn table. Why couldn't they have done that in the first place? *Why?*'

After The Silence I walk back up the line of the burn, stopping to look at the remnants of autumn colour in March Wood. It suddenly strikes me that my feeling for this place has deepened still further in the last twelve months, as the land begins to truly nurture me. The bounty that springs from the soil, whether it be food on our plates, wine in our glasses, or logs on the fire, is on a scale that I have not known till now. This place is more than the kind of home I have lived in before: the implications of it are more complex, and more profound. And with the thought of those who fought 'for King and Country' so fresh in mind, I have to ask myself if the ownership of this place, and the hold it has on me, alters my view of patriotism.

I'm not aware of ever having been patriotic. Partly that is due to being unsure about what it means: is patriotism a love of your own people, or of your own country, or something of both? And partly it is because I have long suspected that an appeal to our patriotic nature may well be bogus. In order to swell the ranks of their standing armies, nations have (right up to the end of the Vietnam War) had to cajole their citizens to enlist and fight for the cause. Over the decades, this took various forms, from Kitchener's pointy finger and the concept of 'duty', through the even more pervasive threat of Nazi occupation of 'England's green and pleasant land', to Vietnam's outright forced draft of, disproportionately, black teenagers.

I am well beyond the age for military service: my children are reaching the point where they also would be too old for war. But my grandchildren are inexorably growing towards that age where they would be ripe for recruitment. I think often about what sort of a world they will inherit when they are old enough to run it, and that brings me back to the depth of my feelings for Garronfield.

I recall a scene from Peter Weir's First World War film *Gallipoli*. The two main characters, unable to join up in their home town in Western Australia, because they are known to be underage, are trekking across the desert towards Perth, lost and dying of thirst. They stumble across a camel trader, who saves their lives with water and directions, and asks them what they're up to. Perplexed to discover there's a war on, and that Australians are fighting Germany in Turkey, he says: "Can't see what it's got to do with us." One of the lads replies: "If we don't stop them, they could end up here". The trader takes a look around at the empty landscape. "And they're welcome to it".

I am a long way from the camel trader's jaundiced viewpoint: there is plenty at stake here in this piece of land. So, what if it came under threat: what if some agency, military or otherwise, were to take it away from me - what if some individual wanted to dispossess me? It is the age-old conundrum: how far will you go to protect what is yours? If you believe, as I do, that it is wrong to kill another human being, then what can be done to counter hostile force?

The argument within myself is being played in the abstract. On a wet November morning this place is not under threat, nor is there anything in the foreseeable future that would make it otherwise. Armed soldiers come marching down the track towards us only in my imagination. But once more it strikes me, with the force of a blow, that this land I own, if not quite synonymous with life itself, has still become my *way of life*, and if you were to take it away from me, you might as well take my life also. The paradox is, I believe this place, for all that it is precious to me, is not worth the loss of one single human life.

Twenty-nine: A Brief History of Farming in Galloway.

At the beginning of December we are visited by two sisters who started out as Patersons, the last occupants of White Croft Cottages. As told previously, their younger brother Cameron Paterson was the first to visit in May 2013, before we had done very much at all on site. His older sisters, Margaret & Sheila, are able to be astonished at the transformation of the place they remembered.

They recall the places they used to play, behind the cottages; where a shed stood and where the hen-run was; they tell us that the front room we'd labelled as the parlour was actually the main bedroom, in which their parents slept - the warmest bedroom in the house, Sheila said. The stone and slate lean-to in the middle of the back wall of the building, the purpose of which had puzzled us, was used as a laundry in their time, with a wash-tub and mangle that needed two hands to turn.

As they reveal more about their time here, connections begin to fall into place. Their father was self-employed, working on local farms, and one of his skills was as a dyker, repairing the Galloway style of walls that I have come to both admire and detest - admiration for their appearance when well-made, but detesting the building or repairing of them - 'upside down walls' I call them, with all the small stone at the base, and big boulders high up: arm-straining, back-breaking work.

Their father had a fine tenor voice, so there had to have been singing here forty years prior to the singing within my own head, that heralded the start of my love for this place. The sisters say that they had a great deal of fun, paddling in - or in Margaret's case, falling into - the burn. Theirs is a fondness for a place that has drawn them back, the last in a 140-year lineage of farming tenants since the cottages were built in 1870.

In south-west Scotland in the eighteenth century the state of agricultural workers' living conditions was, for the most part, dire. Typically, primitive cottages were built of stone bound with clay (the *clay dabbin* method of construction found widely on both sides of the Solway), thatched with heather, sometimes tiled, rarely with a slate roof. There would be a single window if there was one at all. Inside the door the floor would be of clay, and the single room, without a ceiling, would provide accommodation for a farm servant's entire family, and sometimes would also shelter livestock.

All members of the family who were capable of working on the land would be pressed into labour, and there was widespread employment of women in agriculture, a characteristic of the Scottish labour force. Women carried out virtually every task except with horses. They did all kinds of field work, but above all were concerned with milking and cheese-making.

Dairy farming originally had a strong base in Ayrshire, using milk from the eponymous dairy breed of cattle. Dairying then spread into Galloway, where small family farms were particularly suitable for the enterprise. On a typical family farm of less than 150 acres the workforce would consist of one ploughman, one odd man (stockman and delivery driver), a harvester (in season), and four female workers - two milkers, one for the dairy, and one for the house.

Cheesemaking became an important farm industry. Dunlop of Ayrshire was the first commercial producer, supplying cheese to Glasgow and Edinburgh, and subsequently all cheese produced in south-west Scotland was also shipped north under the Dunlop label. Not until the mid nineteenth century did farmers in Galloway begin to market their cheese independently as Cheddar.

In the late eighteenth century a tide of agricultural improvement was underway. There were changes to agricultural practice, and William Craik of Arbigland was a noted proponent, advocating rotation of crops and the use of lime to increase fertility. A part of the Scottish 'improving' policy dictated that only the workforce actually needed for proper farming should be retained. As most permanent agricultural workers were farm servants, accommodation and employment were linked, so that to be without a job was to be homeless. Surplus labour moved away to find work, and surplus cottages resulting from this thinning-out of the workforce were pulled down.

By the 1840s social conscience began to seek improvements to the living conditions of farm servants, and a rash of architect-designed plans for accommodation appeared. An early design for a double cottage that was published in 1844 was very similar to the layout of White Croft Cottages: a lobby inside each front door, leading on one side to the main bedroom, on the other into the kitchen, off which was the pantry, and within this combined space were alcoves for beds. Each room had a window, and outside were the coal-house and privy.

The cottages were probably the first new farmworkers' dwellings to be built in the area for some time, and they were certainly notable: several years after they were built they were still being described as 'the New Cottages at Whitecroft, Southwick'. This is from an advertisement in the local paper, placed by the farmer Mr McKerrow, who leased the land at Whitecroft from Auchenskeoch Estate. One of the cottages was to let, and 'a person supplying a milker' was preferred. A year earlier he had advertised for a ploughman, 'preferably with a wife or sister who could do the milking'.

It is likely that the advert for a ploughman was due to the departure of Alexander and Sarah Buchanan, the first inhabitants of one of the cottages in 1870. We know from John Gillespie's book on the history of the Colvend Coast that the pair lived here for several years, and that two children were born, the second of which died in infancy, a sad occurrence that probably prompted them to move on - not far, though, just to a cottage called Plumjordan, a mile and a half to the south-east.

The couple gaze out of a photograph in Gillespie's book, a posed sitting at the local photographer's studio, presumably. Alexander is standing, dressed in a dark three-piece suit, watch-chain across his belly, right hand resting on the rim of a vase containing a pot-plant which stands on a small table. His expression is serious but not unkindly. He has neatly combed dark hair with a parting on the left, and a full beard and moustache. If he were alive in these times he would pass as a hipster.

His wife, Sarah, is seated at the table, in a dark dress with a high neck, secured with a brooch. At first sight her expression seems serene, but a closer look finds a trace of sadness around her eyes. Her hair is parted in the middle and swept back behind her ears: it may be that her hair is gathered in a bun at her nape. Her left forearm rests on the table, and alongside the hand is an open book - a Bible, perhaps, or a volume of poetry? At any rate, a prop calculated to show that, although farm servants, they had a degree of cultivation belying their station in life.

There is another photograph of Alexander, an older man now, posing alongside his eldest son Jack, who was born four years before the Buchanans moved into White Croft Cottages. They are seated on what seems to be an antique stone bench, in front of a painted woodland backdrop. Jack's father is this time in a check three-piece suit, right thumb in a waistcoat pocket, watch chain secured at the third button from the bottom, and rising to a pocket close to Alex's heart. His beard is white, and there is just a rim of hair visible across the back of his crown. His son, sitting perched a little higher than his father, so seeming taller, is a handsome young man, slim, dark-haired, clean-shaven. He too is in a suit, but the dark jacket and waistcoat are plain, while the trousers seem to be a lighter-coloured corduroy. He wears tightly-laced boots: his father appears to be wearing carpet slippers.

I study Jack's portrait: he looks back at me with a confident gaze, bright-eyed. He's not exactly smiling, but there's a trace of amusement at the corners of his mouth. I would put him in his late twenties, perhaps a bit more, in which case the photograph was taken around the time Alex and Sarah achieved their dream and took on the lease of a 200-acre farm at Tinwald, beyond Dumfries. Jack would work on several farms in the area as ploughman, rising to Farm Manager, before taking on the lease of Clonyard of Southwick, less than a mile south of White Croft Cottages.

He was back in the familiar territory he had explored as a youngster, paddling in Caulkerbush Burn, walking down the fields to Auchenskeoch Mill and Bunghole, and climbing Barn Hill, behind the cottages, to take in the view. Almost a century later it would be Cameron Paterson's turn, about the same age as Jack, to paddle in the burn and walk the fields. Another fifty years, and another crop of youngsters - my grandchildren - run about the place and have the fun that surely was to be found here all along.

Thirty: Midwinter Expeditions.

Barn Hill. Shamed, perhaps, by writing about Jack Buchanan's exploration of the surroundings of White Croft Cottages, and keenly aware that I had never been right to the top of Barn Hill in all the time we had been here, we set out on Christmas morning for a walk around before lunch.

The day is that typical lowland winter weather, low cloud and mist. It is not raining, but the air is so full of moisture it beads our coats and clouds my glasses so that I have to wipe the lenses repeatedly. Wood and stone underfoot are greasy and treacherous.

We climb the slippery post-and-rail at the top corner of Orchard Bank and head straight uphill. I am familiar with this part, because from time to time over the past eight years I have climbed to the same point half-way up the slope to take panoramic photos of our place. Today all is blurred with mist, but I take the photos anyway.

The summit of Barn Hill is crowned with gorse: between the clumps bare rock breaks through surprisingly boggy ground. And what we had suspected, what we had heard by rumour was there, came into view: the long-tumbled stones of a circular sheiling, scarcely a barn, but at least a place to shelter livestock. The old walls now enfold nothing more than a single well-shaped rowan, the husks of berry-clusters still on its outermost branches.

We follow the ridge south-eastwards, and drop down to Alan Carter's modern galvanised stock pens alongside the lonning that crosses from the B-road to Clonyard of Southwick. Into the lane, we turn south. There has evidently been much wind-blow in recent times, and the scraggy remains of ash and thorn have been crudely cut back, the branchwood pushed onto the verges or into the adjacent field. A small plantation on the east side of the lonning contains fine mature conifers, straight and branch-free, perfect for ships' masts.

Further on, tucked into a field-corner, is the tiny roofless cottage known as Bunghole, which was once reputedly the haunt of a character known as Boozy Jenny. Further still, where the burn cuts under the lonning, the derelict buildings of Auchenskeoch Mill, which must have ground corn for the surrounding farms. It had become disused by the Victorian period, but it is recorded that in 1861 the buildings were still inhabited, the mill by Sam Booth and family, and the kiln by Sam Muir the tailor.

A little beyond the ruined mill, astride its own bank beside the lonning, grows the magnificent lonesome pine that rears taller than any other tree around it, and which is the focal point of our view from the cottage to the south-east. The massive lower trunk divides in two some twenty or so feet up, and the twin trunks continue to push upwards and out of the surrounding canopy.

We work back up the Castle Farm fields, passing below the stones of long-abandoned Crossneedle Croft, that was once part of the mill lands. There is no discernible outline of the buildings, just a slew of rough stones, the collapsed walls probably overlaid by stones collected from ploughing the surrounding land. Tucked as it was into the slope below Clonyard Wood, and facing north-east, the croft must have been a cold grim place in winter, with scarcely a drop of sun. It serves to remind us how fortunate we are to have inherited the foresight of the builders who chose the sunny location of White Croft Cottages.

Wild Cat Craig and Doon Hill. This small but steep-sided prominence is visible from the west side of Garronfield. Karen has marked it as a possible place to ride, and so on Boxing Day after lunch we drive down to explore.

The double gates at the entrance are locked, but there is a way through to one side. We take a clockwise direction along what promises to be a circular route. The track zigzags, gaining height, until it levels off to strike along the woodland edge behind Dunmuck Farm. Three horses stare up at us as we pass: a dog barks in the yard. Apart from the dog there is silence, nothing moves on the road below, and no birds call out.

The track turns round the sharp north end of Doon Hill and heads south-eastwards, with just a fringe of trees to the left, offering occasional glimpses into the fields below. A raven flies up off Redbank Hill, across the narrow valley, and plunges out of the air, twisting and calling, no doubt showing-off to his mate perched unseen in a tree on the slope opposite. Further along Doon Loch comes into view, a heart-shaped lochan, and when we are looking straight down on its quiet waters we can see that in an adjacent field, down from the point of the heart, there is a well-built circular enclosure in the middle of the field, trees growing within it.

There is a story here, there has to be. The lochan does not appear to have accidentally shaped itself like a heart, and there would seem to be no agricultural purpose for the round enclosure. I sense there may well be love and loss involved in the making of this landscape. The strange thing is that although the enclosure is clearly old, and of a decent size, it does not feature on the OS map.

At the southern end of the hill the track turns towards its beginning, and passes down through a mature beech wood, beneath the crags that presumably were once the haunt of wild cats. Someone takes care of this place: branches overhanging the track have been pruned back properly to each trunk, and the prunings laid out neatly on the ground between the trees. And in a small field close to the entrance to the wood, three ancient oaks of huge girth.

Portling and Port O'Warren. Once before we had tried to walk the section of the Solway coast from Portling around the headland of Blackneuk Craigs to the natural harbour of Port O'Warren, but it was high tide, and the only way led across the tilting beds of fissured rock above the tideline. Sky the dog was still with us, but increasingly infirm, and the route was too difficult for him - though he tried gamely, as ever - and we turned back.

This time, New Year's Eve, the tide is well out. Oystercatchers and a few curlew work the mudflats, and further out a fringe of small waders are busy at the sea's edge, too far away to identify them. Beyond there is a blank shroud of grey across the Firth: the Lakeland fells and even the turbines of Robin Rigg are invisible. But to the south-west a golden curtain of sunbeams pours through a slot in the clouds.

We round the craigs and enter the spectacular harbour, lined with rock walls, the mudflats leading to a narrow pebble beach, where a driftwood bonfire is being prepared for the coming night. To the south-west are the inhospitable cliffs of Millstone Loup, Gillis Craig, and Cow's Snout, but Port O'Warren offers sanctuary, which was evidently made use of by smugglers and their contraband, as well, perhaps, as legitimate cargoes. Behind the beach, almost down to the tideline, a cluster of houses sit on ledges cut into the steep slope. There is a bustle of activity: the bonfire-building, provisions being carried from 4x4s to front doors, and a sense of excitement at what lies ahead.

We strain our leg-muscles on the steep click above the houses: the concrete access road would be impassable to any vehicle when icy, and it reminds us that 'warren' might well be a corruption of 'garron', those sturdy, sure-footed ponies that would have been essential to haul loads up out of the cove, for smuggler and law-abiding seafarer alike.

Round Fell. Our first outing of the New Year, and we take the Gliding Club track through clearfell and round the back of the plantation on Round Fell. This hill looks down favourably on Garronfield: it and neighbouring Maidenpap form our skyline to the north, and Round Fell in particular appears enticing in afternoon sunlight, with the warm glow of bracken and heather, and a topknot of conifers. While we are enjoying a break from all that happens on our parcel of land, it is an opportunity to explore another prominent part of our landscape.

My idea is to follow the forestry track to the point where it is closest to the summit of the fell, and then find a way, along a rack or a fire-break, perhaps, to a point where we might be able to look down on our place. But it proves impossible: all potential ways through are choked with windblow. We find one open rack that leads into a fire-break, but that comes to a dead end. On our way back out, though there has been not one sign of life in the forest, a snipe does a vertical take-off in front of us and whirls away above the tree-tops.

Back down at the forestry gate we encounter a Gliding Club member struggling with an unwieldy padlock. Until recently this gate has not been locked, and Karen asks him why it is now. He shrugs: "It's a new broom - yet another Forestry Commission land manager, with his own ideas of how things should be. We've had four different ones in the last four or five years. Heigh ho."

Mersehead. The following day we drive down to Mersehead Nature Reserve before dusk, to look for geese. We walk out to the furthest hide, and there are some barnacle geese in the fields alongside the track, their presence betrayed by one dark head lifted at all times, alert for danger while the others feed.

When we are among the trees of Mersehead Plantation, approaching the hide, a skein of barnacle geese fly over, not the concentration we had hoped for, but a welcome sight, and a chance to hear their distinctive barking call close to. I don't think this species overfly Garronfield, except perhaps on their migration: the geese we hear overhead regularly throughout winter seem to be mainly pinkfeet.

The hide overlooks a lagoon, created during the reflooding of Mersehead to provide habitat for wildfowl, the barnacle geese in particular, since the entire population from Spitzbergen (Svalbard), above the Arctic Circle, overwinter on the Solway coast. On the lagoon we watch wigeon, pintail, shoveler, mallard. A pair of Canada geese preen themselves in one corner of the pool.

Then - it *is* pure gift, as Sara Maitland said about the hen harrier that hunted along the burn by her cottage - right in front of me the brown slicked head of an otter emerges from the water, a vee of ripples following behind. The animal swims to the bank the hide sits on, and emerges briefly, before some noise or other we make in our excitement alarms it, and it shoots away back the way it had come, leaving a cloud-trail of mud hanging in the water.

Lord Maxwell's Cave. Our last expedition of the winter break, and I want to look for two enigmatic points marked on the OS map - Grey Ox and Lord Maxwell's Cave. These lie in the rough hilly terrain between Drumstinchall and Clawbelly Hill, a couple of miles to the north-west of Garronfield.

We follow the forestry track through Isles Wood which would seem to bring us closest to our goal, from the north side. On the map it looks as though the track dwindles to a path before emerging from the forest on to open ground. But there is no open ground: in the score of years since the map was last revised, conifers have self-seeded themselves across a broad swathe of the terrain. Their colonisation is haphazard, and there are openings we make our way through with difficulty.

We climb to a knoll which gives us a view of the surroundings. The cave should be somewhere below us and off to the left, but it is clear that to search for it would be quite a task, with the trees that shouldn't be there, according to the map, closing off any chance of seeing the place from a distance. We would have to be lucky, and stumble over, or perhaps into it. We decide to leave Lord Maxwell in peace, and head back northwards to find Grey Ox.

I had thought that this was likely to be a large boulder, or an outcrop of rock that in some way resembles the eponymous animal, yet once again the self-sown trees thwart any chance of spotting anything from a distance. However, this time we do come across a boulder, ringed by conifers, that seems to be in the precise location of Grey Ox on the map. Karen is unimpressed: it is scarcely huge, and nothing about it is suggestive of a cow. I remind her that in the past people who had cause to venture into wild terrain would often name distinctive features to provide others with a guide for navigation.

But she isn't convinced, and we carry on northwards, aiming to come across a faint path across the hill that links the forest track we had come along with a parallel one to the east. Some months before, Karen and her friend and riding companion Vyv had found this route, which was a bit difficult for the ponies, yet still passable. But somehow we fail to find the path, and begin to descend through ever-thicker clusters of trees, to the edge of the plantation proper. Without an opening we can go no further.

We decide to work across to our left, hoping to be able to drop down onto the forest track, or find the path across the hill. We burst through a number of tree-barriers, and eventually come across a vague opening where some thinning has taken place in the past - a fire-break presumably, but done so long ago that yet again self-sown trees have grown up to almost obliterate the break. It is difficult, ankle-twisting progress, not knowing where we will emerge, or whether we will have to retrace our steps.

But now the fire-break begins to widen, and we zigzag around a few more trees before emerging on top of a steep bank above the welcome sight of a forestry track, which Karen recognises as the one running parallel to the one we'd taken earlier. With a great sense of relief we head back to where the car is parked at the forest entrance.

All afternoon it has seemed as though we are the only living creatures in the forest. Just once a wren flittered across the fire-break in front of us, and when we are nearly back at the car a robin darts out of a small tree on to the track, picks at something on the ground, then flies away from us. Otherwise, there is silence and stillness all around. But although we see nothing, no doubt pairs of night-vision eyes track our stumbling, crashing progress, from within dark portals on either hand.

Thirty-one: The Food Bank.

As soon as hoar frost has wilted off the grass in the glare of a late January sun, the gulls appear. This morning I do not spot them flying in, they are all at once just *there*, peppering the slope of Barn Hill behind the cottage. There are more than two hundred, perhaps three, mostly common gulls with a scattering of black-headed gulls. By contrast to their raucous noise down at the shore, they are eerily silent. A few latecomers drift in, but the flock moves as one, marching across the slope to the dyke dividing Alan Carter's fields, then lifting and flying back to their starting point.

Unless they have taken a fancy to sheep droppings as a food source, I can only think that this particular south-west facing slope is rich in invertebrates - literally a 'food bank'. Once or twice I have seen a similar congregation across on New Farm's steep field, but it is a daily occurrence now on this side. I strive to think when it became so, because gulls appear on the slope at certain times throughout the year. This regularity, and this concentration of them, seems to have begun only in the last weeks of the old year.

It is not just the gulls that favour the slope: the sociable crows - jackdaws and rooks - seem to like it also, though they appear to have become displaced by the gulls. Back in the autumn it was common to see a fair-sized crowd of crows wheeling and tumbling above Barn Hill, then floating down to land on the sward.

Outside the breeding season curlews also visit, though not in such great numbers. It will be a small gang, not more than a dozen, that glides in to land far up the slope, further up than gulls or crows, so that they are out of sight from below, or just their heads appearing above the horizon. A curlew's call always makes me look for the source: the plaintive whistle, which can carry a fair distance, seems to me the loneliest and wildest of any bird call. It speaks to me of wide and bleak uplands, of the openest of open spaces, and I feel honoured that on occasion they choose to visit our less bleak and more enclosed space.

Some winters we do not see any of the Scandinavian thrushes, fieldfare and redwing. But on the two or three occasions when they have come, it is this same slope they visit. One year there was such a concentration of them that they spilled over the fence, out of the field and onto the track alongside Spring Wood. There were puddles of rainwater on the track, and some birds took the opportunity of having a bath, but curiously it was only the redwings that did so.

Sometimes starlings visit the 'food bank', a close-knit smudge of them working their way across, fifty of them at most. But from late autumn onwards they do not stop off to feed: before dusk they are on their way to roost, most probably in the reedbeds at Mersehead, and as they fly over our patch, more often than not they perform the extraordinary aerial ballet known as a murmuration. Ours is not the mass display of tens of thousands of birds, but a smaller event, perhaps two or three hundred. Nevertheless, it seems as though when they are over Paradise Meadow something inspires them to fly up in a column, then form hoops and figures-of-eight: a brief dance before they continue on their way.

Low sunlight across the slope of Barn Hill reveals parallel shallow depressions running diagonally downhill. They are too far apart to be the rigs of ploughing, and it would be a steep bank to plough anyway, so I suspect these are the lines of stone drains, formed by digging a trench and filling it with stone gathered off the land. In fact where we have dug into the bank at the back of the cottage, there are patches of small stone that we have cut across and exposed, just under the topsoil and the sward.

The lines speak of a time when great care was taken to improve the land, to keep it well-drained and as productive as possible. Nowadays the four or five fields behind us are farmed more or less 'ranch style': the dykes separating the fields are tumble-down (except the one around the silage field, which is maintained in reasonable condition to keep livestock out), and the stock graze the whole hill as one pasture. Sporadically artificial fertiliser gets applied, and sometimes the pasture is topped when thistles are getting out of hand. But that is all the attention this block of land gets.

In contrast to this low-key attention, the birds give the food bank all of theirs. I wonder if the riches to be found there are the legacy still of the care that was taken of the land many generations ago, the drainage helping to improve the soil and produce an abundance of invertebrates.

That legacy must of course be combined with the simple fact of the slope's aspect: it faces south-west, and apart from the early morning it basks in whatever sun the day brings, until evening. In winter the summit of Barn Hill is still lit when all else around has sunk into the gloaming.

Thirty-two: Living Room.

The sash window in the living room has four panes. At nightfall on a mid-February day the bottom two panes are completely dark: I know what is out there but I can't see it. The top left pane is also dark, except for the lights of the latest building at New Farm, which is used for lambing at this time of year, and is continuously lit, making it look like some alien craft perched on the horizon.

The top right pane holds the very last light of day, a faint green glow that delineates the skyline, which rises left to right along the ridge, up to the edge of Bennel Wood. In this minimal light the trees seem like an enormous black wave overtoppling the ridge, threatening to engulf us. I wait until the final glimmer has drained from the sky before drawing the curtains.

I turn back into the living room and begin to light the open fire. This is a time of reflection, after the day's work is done, with the evening opening up like an enticing box full of treasures, and tonight I find myself thinking about the term 'living room', and why I have preferred it to the equally appropriate 'sitting room'.

Part of the reason comes from a long way back. I recall reading, in my early twenties, a book about the life of a country vet, in which he recounts an evening visit to a lonely farm that was run by four unmarried brothers, all Methodists. As the vet went to knock on the farmhouse door, he passed the lit living room window, which was curtainless, and couldn't help but glance inside. There were the brothers, sitting in front of the fire, in a row on a long wooden bench seat. They didn't talk, they didn't read, they just *sat*.

As I embarked upon marriage and family life, I had a sequence of front rooms, and each contained comfortable chairs, certainly, but there was always much more going on than just simply sitting. There was TV and music, games to play, stories to be told: far from being a vacuum of time after work, this was *living*, and so the room where so many relaxing and enjoyable events took place was always the living room.

The naming of rooms took on a new importance as White Croft Cottage came into being, just as important as naming places on the land. There were, of course, drawings for the purpose of the planning application and so on, the rooms being assigned names to identify their purpose to third parties examining the plans - not necessarily representing the actual usage the spaces would have later on.

The living room was marked as such on the plans, but the smaller room leading off it was named 'dining room', though I can't recall any stage at which we thought we might eat in it. So it finally became Karen's office, and temporary bedroom at times of full house. At the other end of the cottage, the room just in from the side door was named as 'utility room', a term I dislike. I had put this down to a feeling that it was a modern Americanism, but to my surprise I have found out the term is an old British export, being first mentioned here in 1760, becoming the ubiquitous terminology in the US in the twentieth century, displacing the older 'laundry'. In Britain the utility room would previously have been known as 'the scullery'.

Out of this tangle I pull the laundry thread, simply because the room is where the washing machine sits, and where washed clothes are air-dried when they can't be dried on the line outside. I don't do the ironing in there, however, which would be the completion of the laundry cycle, and there is a small chest freezer purring away in the corner. Nevertheless, the room is called the laundry, and probably always will be.

The room where I sit to write is marked on the plans as 'study', a place to be studious. But somehow the word seems inadequate to encompass the uses of this space. One wall is lined with most of my books, so it could be a library, and I do my paperwork in here, so it acts as an office. And on occasion the room is a place of refuge, an escape from what is going on elsewhere in the house, so perhaps the term 'den' is the most suitable and all-encompassing.

Then there is the room that appeared on the plans as 'bedroom 2', with attached shower and toilet. This was where we slept when we moved into the cottage, before the main bedroom and bathroom were finished. Now we most often call it the spare bedroom, which is literally true, as no-one sleeps in it regularly. But 'spare' suggests to me a waste of space, a disuse I am uncomfortable with. So I am trying, without complete success, to call it the guest bedroom, for it is simply waiting for the next guests to arrive.

In the north corner of the kitchen is a small unheated store room. There are thick sandstone slabs on the floor along one wall, and above these are more slabs on a timber frame, at worktop height. The rest of the floor is covered with reclaimed quarry tiles. At head height an L-shape of broad shelving completes the fitting-out of this space.

It is, of course, a food store, something that seems to be omitted from many modern houses, but we had the room to construct it in traditional style. I grew up with one such, in a Lake District cottage, where the store room on the north side of the kitchen had a slate flagged floor, and a single huge slate slab shelf that was always cool to the touch, even in the hottest summer. I can recall the room vividly: what I can't quite be sure of is what we called it.

Somewhere in a back corner of consciousness I seem to hear my mother saying: "Go and put that away in the pantry, Rob". I have tried substituting the word 'larder', but it doesn't sound right, so I guess that store room was probably known as the pantry.

The terms are interchangeable to some extent: they both come to English from Latin, via French. But larder (*lardier*, Fr; *laridum*, L) has, as might be expected, meaty connotations, whereas pantry (*paneterie*, Fr; *panis*, L) is the place to store bread, and, by association, other foodstuffs. As we don't keep meat in our store, and given the antecedent of the room I grew up with, I am more than comfortable to name the cool room in the corner of our kitchen, the pantry.

This rambling tour of the interior placenames of White Croft Cottage brings me back to the living room, and as I stand in the doorway I can see afresh how comfortably set out it is for relaxation, with my armchair and Karen's couch arranged in front of the fire, and a second settee along the wall to my right. When we have any more than two guests we can rearrange the chairs to form a wide arc around the hearth.

This happens most often when my family descend upon us, sometimes nearly all of the tribe together, filling the house with their bright clamour and excess energy. Then the living room does come alive, becoming a den, a place to hide and seek, to play games on the carpet, to toast marshmallows in the flames of the fire.

But when they have all climbed back into their cars and bumped away down the track, we wave until they are out of sight, then turn back indoors and rearrange the chairs into a closer intimacy, because most of the time - almost every evening - the room is a place to sit and relax, listen to music, watch a film, read a book, sometimes even have a conversation. There is plenty of sitting room in the living room.

Thirty-three: March.

Alec is coming over tomorrow to help me with coppice work. He has spent two Saturdays here in February, learning about the cycle of willow production from planting to harvest, as he is studying horticulture at Threave Gardens in Castle Douglas, which is owned and run by the National Trust for Scotland. Alec is the younger son of Margaret Paterson, one of the last family to live at White Croft Cottages, and he accompanied her when she paid a visit in December last year, along with his aunt Sheila. It emerged then that he was interested in all that we are doing here, and I'd offered for him to gain some work experience in the spring, which he'd accepted with enthusiasm.

I like him a lot: he is intelligent, practical, polite, thoughtful. He likes to work, and he is in his element here, among the willow stools, making cuttings, planting, harvesting grown rods, sorting through them to pick out the best. He learns quickly, and picks up on the traditional ways, and the terms used to describe the processes. Tomorrow we are going to work through the already-felled coppice in Spring Wood. He has some experience of brush clearance, when he worked for a time with the Galloway Wildlife Trust, but this will be a more deliberate operation, extracting useable timber, and utilising the brash to deter deer from browsing the regrowth.

There is a poignancy in this. At twenty-three, Alec is just one year older than my eldest grandson Jordan would have been if he were still with us. Like Alec, Jordan was interested in the outdoor things that I was engaged in, and in the autumn before he died I asked him if he would help me plant trees here in the following spring, a prospect he obviously looked forward to. He broke the contract when he took his own life a short while later: whether he meant to succeed, no-one can be sure. And so it was that he was the only one missing when my family spent a February Saturday planting trees, including the holly grove that is his memorial here.

Alec turns up at ten o'clock, right on time as usual. I am not really fanatical about time-keeping, but it has been observed that people who are habitually late are generally more cheerful than those that wait for them.

We get to work straight away. He listens carefully while I explain the task: the trees have been felled already, their tops pointing uphill, and I will go in first with the chainsaw and cut the poles into roughly five-foot lengths, while he will follow after, trimming off side-branches, and spreading the brash downhill to form continuous cover around the cut stools. Although he will know the safety drill from previous work experience, I deliver it anyway. I say to him, "If you do decide you're going to cut off one of your fingers, at least I will be able to say, hand on heart, *well - I did tell him...*"

There's a choice to be made, between William Hook, my trusty and ancient billhook, and a Brazilian blade, which might be called a machete, that I bought many years ago from Banks' Ironmongers in Cockermouth, the fabulously old-fashioned establishment where you could buy a single nail if that was all you wanted. I'd gone in to buy a billhook, but there were none. Mr Banks told me that he'd just taken delivery of some South American blades, and showed me one. I liked the feel of it, and bought it. As I was on my way out of the shop, Mr Banks - who was a local JP - said: "Hang on, I think I'd better wrap that for you, or you might end up in court in front of me for carrying an offensive weapon in public, which would be embarrassing for both of us."

Alec opts for the machete, soon getting the feel of it: he says it has a nice *heft* - exactly the word I would use to describe its action. We talk about the traditional terms for what we are doing: he knows some of them, but others are new. Trimming side-branches is *snedding*; we are *working up* the cut coppice, and the cut area can be called a *coup* or *coupe* (the term Alec knows), a *panel*, a *cant*, or just simply a *cut*. The terms are ancient, their exact definition no longer in the OED or Chambers, only by inference, but essentially they refer to a cut area of coppice or a division of a larger area.

In pre-industrial times so much use would have been made of coppice products: bark for tanning; small poles for bobbins, hurdles, light construction; thicker poles for building, boat-making, and firewood; charcoal-making - the list goes on and on. Even the brash that we are spreading so prodigally across the ground would have been cut up and bundled into *fascines* (for stabilising riverbanks and steep slopes) and smaller *faggots* (for heating bread-ovens). We speculate about how they managed to prevent browsing by deer if the coppice floor was bare of brash.

The day passes in pleasant manner: the weather is fine, just a few spots of rain before lunch, and the wind is light and coming from the right direction to help us as we heave the brash over. Frogs leap from underfoot; there are badger footprints to see on the track; Alec is once again, as he admits, in his element. The varied nature of estate work would be his idea of the perfect job, and over lunch Karen suggests some possible avenues to explore for when his course at Threave ends in August.

I will get the coppice work finished over the next few days, so this is his last visit for the time being. As he makes to leave he is fulsome in his gratitude for having had the opportunity to learn new things, and I am equally fulsome in my thanks for his excellent help. At the last minute I decide to tell Alec about Jordan: his empathy and compassion are immediate, genuine, and very touching.

Three years into establishing coppice the results are apparent and encouraging. In the top triangle of Spring Wood in particular, the three coups are adjacent, the stages of regrowth easily compared. In the first season of regrowth, willow naturally takes the lead, putting on at least eight to ten feet, with alder and ash some way behind. In the second season the willow poles gain in girth, but not so much in height, whereas the other species gain in both girth and height.

One concern about having a high proportion of vigorous willow was that it might shade out slower-growing species. It may be this is not going to be the case, indeed the willow may well be beneficial as a nurse species, providing shelter and stimulating competition, but if it were to become over-dominant, it might become necessary to coppice the willow only at three years and give the rest more growing-room.

Each time I have stood alongside a worked-up area of coppice I have felt a sense of pride and satisfaction at the result. Hard on the heels of this feeling comes a sobering thought - it may be at least five years, possibly seven, before I set foot in there again with my chainsaw. I will be in my late sixties, or early seventies.

The first time I realised this, it was rather unsettling. We had merrily spoken previously of a seven-year rotation for the coppice, without my making a direct connection to the rest of my life, but it became clear that my time to come would be parcelled out in coups, panels, cants - just as the woods themselves will be. I have grown to like the knowledge of this: it is another way in which I can feel at one with this place.

In its last week, March lives up to the old adage and departs lamb-like, with calm, settled weather, often overcast, but dry. This in contrast to the boisterous days earlier in the month, with incessant gales and strong gusts and lashing rain. By coincidence there are lambs on the skyline of Barn Hill, lying in the sun while their mothers graze.

Birds are pairing up and there is much singing. The first trees, in sheltered spots, are coming into leaf. The imperative of spring is once more surging up and out through every shoot and branch.

Thirty-four: Easter Rising.

The holiday is just about as late in the year as it can be, as determined by the arcane calculation of ecclesiastical months and full moons, perhaps the established church's equivalent of football's offside rule.

The weather, which has been governed for some time by chilly easterlies, warms up considerably, with the Met Office getting excited about the prospect of record temperatures. As the warmth increases, Garronfield once again undergoes the rapid uprising of green growth and the upcountry movement of our summer visitors.

On Good Friday the first willow warbler sings in Spring Wood, concealed among fresh bright foliage. On the Saturday a swallow wings past us, heading north, and a cuckoo calls from Bennel Wood, though Karen thinks she heard him earlier in the week. After lunch Ben and Pete and Amy turn up with a gaggle of seven youngsters, bringing their sweet brand of chaos, a mini-revolution amidst our normally ordered and measured life.

Later we pick a variety of leaves - mustard, rocket, pak choi, broccoli, and more - to make a salad for our evening meal. There are the first four asparagus spears thrusting upwards in the allotment: not enough for twelve of us to share, so we decide to be selfish and keep them for when the family have departed, which they do on Sunday afternoon. Amy tidies the kitchen, which had been transformed into a craft workshop, a riot of paper scraps, pens, sellotape, and staples. Mollie manages to leave her Easter basket behind, and it sits in the middle of the big table. Silence laps back in around us like a tide.

On Easter Monday the first of our own swallows arrives, flitting around the outbuildings, recalling its birthplace and prospecting for a new nestsite.

The only thing that is not rising this Easter is our spring, which has not been visible on the surface for more than a week, and becomes unusable on the Saturday morning. I lift the lid on the concrete ring that houses our water supply, pull the outlet pipe out of the remains of springwater, and thread it into the 200-litre plastic barrel that is the end part of our rainwater-capture system, and which sits permanently inside the concrete ring as our backup in times of drought.

When we first tapped into the spring, in August 2012, we thought it to be a year-round supply. It soon became apparent this would not be so. In the last week of June the following year, after an extended period of dry weather, the spring began to dry up, and was unusable by the end of the month. Our first rainwater capture set-up was off the workshop roof: I put guttering along both sides and across the back gable, and ran a horizontal pipe from the bottom of the downspout to the plastic barrel installed inside the concrete ring.

It got its first use a few days later, when a thunderstorm hit Garronfield, and I watched with satisfaction as water cascaded from the end of the pipe into the barrel. I sheltered in the workshop while the rain hammered on the roof and lightning flashed, until after about ten minutes there was an ominous crack that was not thunder, and I rushed outside to find that the brackets I had used to convey the guttering across the back gable had not been strong enough for the weight of water, and this section had come adrift. The guttering along the side nearest the barrel was collecting rainwater, but half of it was pouring uselessly out of the broken end. By the time I had fixed this, the thunderstorm had passed. The barrel was scarcely half-full.

We made plans to improve our water storage by digging in a second concrete ring on the bank above the spring, to be filled in times of plenty, and gravity-fed into the barrel in times of drought. The spring returned after serious rainfall in early September, ten weeks after it dried up. We hoped that the unusually dry spring and summer were the cause, an abnormality rather than otherwise.

In May 2014 we installed the first phase of a more sophisticated rainwater capture system. Karen and her Dad put up 15 metres of guttering along the back of the barn that we'd built the previous autumn, while I constructed a platform at the end of the building, across from the polytunnel, for two thousand-litre tanks. The pipework from these led to a tap in the polytunnel, and the barrel could be filled by hose from this tap. It was a temporary arrangement, and we planned to plumb the system in so that the barrel would fill automatically, but that would have to wait until we dug a trench across the yard.

The set-up required an effort of memory: to remember the hose was on, and turn it off before the barrel overflowed. There were many occasions when I would suddenly leap to my feet, cursing, and race outside to turn off the tap. Many hundreds of litres of precious rainwater got wasted because of forgetfulness. The new system came onstream on July 3rd, and was in use till October 9th.

The following year was an anomaly. As we were coming to expect, the spring dried up on July 1st, but heavy rain three days later restored it, and it continued bubbling up until October 9th, when it dried, returning on November 4th. This autumnal failure of the spring was a new and unexpected phenomenon.

Through the first half of 2016 one of our main jobs was laying the floors throughout the cottage. We were using a breathable system, with a lime-based screed, and this was very demanding of water - a full day's work would consume around 300 litres. While the spring was running this wasn't a problem, but it stopped before the end of May. The weather was abnormally hot and dry. Every morning the Met Office promised downpours, but each day the black clouds passed to west and east. Someone was getting drenched, but it wasn't us.

By mid-June our rainwater supply was looking ominously low, and still there was no sign of rain. Crisis point arrived, and I began ringing around to try and get a bowser. It turned out there were plenty of fuel bowsers available, but none for drinking water. Eventually Stevie of Colin Dempster Plant Hire came up with the solution. He called in a favour from Scottish Water (who just an hour earlier had refused to supply us with emergency water, as we weren't a customer) and the next day we had a bright blue thousand-litre SW bowser parked next to the spring. Stevie was familiar with our place, having worked as a contractor on the farms round about, and his boss Colin had lived in one of the cottages for a time, probably in the late 1950s or early 60s.

The crisis did not last long: four days after the bowser arrived the rains came, and delivered more than four thousand litres over the course of a few hours. We got by on the rainwater supply until August 20th, when the spring returned.

2017 was another anomalous year. The spring dried on May 7th, a fair bit earlier than previous years, but then heavy rain restored it a week later, though with little pressure. When the spring is working properly it bubbles out of the ground energetically. I wasn't surprised when it only lasted four days. But then normal changeable weather brought the spring back on June 6th, and it was with us for the rest of the year. In September it did die right down, but recovered.

The crisis of 2018 was of a different order to anything previously. May was a fine dry month with very little rain, and the spring diminished, as expected, but quite slowly. It dried on June 2nd: heavy overnight rain topped up the rainwater system mid-month, but there was nothing more. At the end of June we had about fifteen days-worth of water left, but no forecast of rain within that time.

We knew that we needed more storage capacity, so ordered a 3000-litre tank as a backup. Water for the ponies was a priority, so we bought a bilge pump, dug a sump in the burn just above Irish Bridge, and pumped water every day into a small trough in the field. Our friendly neighbours, Pat and Stewart, loaned us the use of their 250-litre bowser, to fill from their own tap, although since they themselves were having some supply problems we didn't avail ourselves of their generosity too much. Every other day throughout July Karen took six 25-litre containers in the back of her jeep, and filled them at the council depot next to her office. If we were careful, we could manage on about 100 litres of water a day. On the 28th heavy rain filled the main tanks and I was able to part-fill the new tank, which was temporarily sited next to the spring, and would be moved to its permanent position when our water supply was back to normal. Crisis over. The spring eventually returned on September 11th.

In the run-up to this Easter I made a pad of concrete blocks for the new water tank, behind the tackroom, a short distance from the main tanks. I drained the new one - it was nearly full - in order to move it, which we did on the Good Friday. The spring dried completely the following day.

We have an empty tank awaiting good rainfall, and I have an almost obsessive need to watch every weather forecast. Frontal systems are predicted, but they are set to come from the east, an unusual direction, and we are in the rainshadow of Criffel if that is where the weather is coming from. We say to ourselves, in wonder and exasperation: "This is south-west Scotland - *it's supposed to be wet...*"

It is tempting to think of the spring as a living creature: unpredictable, mysterious in its comings and goings. In recent years I have spoken to other people with appropriate experience, and the consensus of advice is: if you have a spring, don't tamper with it, for if you do it may well disappear for good. Every day now as I pass our spring and glance at the dust-dry channel around the concrete ring, that just a couple of weeks ago was running brimful with clear pure water, I have to wonder if our supply of water from out of the ground is gone for ever.

Thirty-five: The Oak and the Ash.

Each spring we observe the order in which our trees come into leaf. Elder is always first, almost sneakily: it seems as though one day there is no greenery, the next the tree is in full leaf. Perhaps this illusion is due to there not being a strong contrast between the colour of the leaves and the greenish bark of the branches. The pioneer species - willow and birch - are early leafers, and the 'shrub' species - hawthorn, rowan, hazel - are likewise early, even where they are growing in the open, rather than sheltered by taller trees.

We take a particular interest in the leafing order of our oak and ash trees, to see if there is any validity in the old saying:

If the oak's before the ash,

Then you'll only get a splash;

If the ash precedes the oak,

Then you may expect a soak.

That is a rather dusty version, from *Brewer's Dictionary of Phrase & Fable*. I am sure that the rhyme comes from country lore rather than an academic source, in which case the language would have been vernacular, perhaps more like 'Oak before ash, in for a splash; ash before oak, in for a soak'.

In this year, 2019, the oak is far ahead of the ash. This is true of our own young trees, but even more markedly in the prominent trees in Castle Farm's field, just over the March dyke. Here a mature oak and ash grow close together, branches touching, almost entwined. In late May the ash is barely in leaf, the oak is luxuriant.

Historical data suggests that ash leafed first 30% of the time, yet according to the Woodland Trust the tree has beaten the oak on only a handful of occasions in the past half-century. Apparently oak responds to temperature, while ash responds to lengthening daylight, so that warmer springs favour oak. It is calculated that every one degree rise in temperature gives oak a four-day advantage over ash. Climate change will have a significant effect on the species make-up of our native woodlands, if ash is predominantly outshaded by oak.

It is not just climate change affecting woodlands: disease is playing its part. Whereas in previous years oak and ash have generally been neck and neck in coming into leaf, with individual trees in particular locations bucking any sort of trend, this year in both March and Spring Wood the oak was emphatically first to come into leaf. And as we waited for the black calf-foot buds of ash to burst we saw that ash dieback has now taken firm hold here.

The fungal disease that causes dieback has been in Britain since 2012, brought on plants imported from Europe. What had first been seen in Poland in 1992 spread westwards and northwards through the EU and Scandinavia. The fungus was only identified in 2006, and its genome not sequenced until December 2012. There is increasing information about the disease, but little consistency.

Evidence from Poland, where dieback has been studied for longest, suggests that 15-20% of ash trees survive, showing no symptoms. Elsewhere, more doom-laden scenarios point to almost total extinction. Unlike elm disease, ash dieback is chronic but not necessarily lethal. It is particularly destructive of young plants, but older trees can survive attacks, though they may succumb after several years of infection.

All attempts to control the spread of the disease have failed. Destruction of trees in infected areas has proved counterproductive, as it removes disease-resistant trees as well as dying ones. A Danish study has found that genetic variations in ash populations affects susceptibility, and research continues. The attempted breeding of disease-resistant trees is ongoing, but it will take decades to repopulate Europe.

Our ash trees had come from a nursery in Northumberland, a year before dieback appeared in the country, and they showed no signs of disease until 2017, so it is reasonable to suppose that the fungal spores have come from the surrounding area. When Karen is out and about with her work she is now noticing dieback in hedgerow trees alongside the roads surrounding us.

It seems almost wrong to say that anything in nature is downright ugly: even the bright orange patches of rust fungus on the leaves of my basketmaking willow have their own beauty. But when it comes to dieback, there is ugliness, in the yellow gangrenous necrosis that taints the bark. Some strongly-growing ash appear unaffected, but weaker trees have succumbed, often in clusters, seeming like groves of skeletons, holding up pale leafless arms, motionless in the very wind that brought the disease.

For all that, what we are seeing *is* dieback, rather than death. A few trees look to have died completely, but those that are affected have some degree of new growth, which is sometimes extremely vigorous, as though the trees are putting in a mammoth effort to overcome the debilitating effects they suffer. Their crowns may be dead, but there is life yet in the lower trunks, and some appear to be almost self-coppicing - relinquishing vitality in the main part of the tree, whilst putting out strong shoots from the base of the trunk.

Taking a lead from this phenomenon, our intention is to mark all severely affected trees in the autumn, before leaf-fall, and coppice them through the winter. It will be experimental, to see if the invigoration of coppicing has a beneficial effect against dieback as the trees regrow. And where a wide area is affected, we will interplant with alder and aspen, so that if the coppiced ash do die, there will be something else to take over from them.

Thirty-six: August.

The extended spell of mostly dry weather, which left us without our spring at Easter, comes to a sudden and dramatic end in the first week of the month. Thunderstorms loom over, slow-moving, heavy grey walls shading up to black. The rain they loose is almost vertical. Then after the rain the storms fetch a bout of wind, that flexes the willows so that the leaves reveal their paler undersides.

They are showing their underskirts, as I have said previously, and it seems a fitting image. But then I wonder: does anyone wear underskirts these days? What about Grayson Perry? I think he may well do.

A couple of days later, a different rain, in still weather, a light but persistent drizzle that bends the same willows, as though they are weary, rain-sodden leaves all pointing to the earth. Our spring returns to us on the ninth day.

I have a new companion in the woodshed: a robin. There is no way to be sure, but I have a feeling that it is one of the youngsters from this year's hatching, sporting newly-acquired adult plumage. It is a curious thing, but every year we see a pair of robins in the first part of March, then not a sign of them at all. The next robin to appear is always a juvenile, and this year it was the last week of July when I spotted the first one.

The bird books say that juveniles disperse from their birthplace, but I am not convinced that has happened in this case. What I am seeing is new behaviour: a close and bold proximity to me as I work, far more intense than I have ever before encountered in a robin. This is what leads me to think that this is a young bird, that so far has had no cause for alarm.

On wet days I have been processing the scrap willow that is unsuitable for basketmaking, cutting it into kindling and bundling it to make faggots. On the first day the robin was already in residence in the woodshed, and my arrival was greeted in a very chip-chip-chippy manner, but the bird settled down after a time, and began to utter brief snatches of song in between ticking me off.

If it was spring, the singing would lead me to think the bird was most likely a male, but in autumn both sexes defend territories, and both sing their winter song, a sweet fluting that is one of the few birdsongs to be heard at this time of year. This robin sings so close to me I can see the voicebox swelling with sound below its open beak.

The bundles of willow are leaning against the shed wall, and the robin frequently scuttles along the space behind them. As I lift each one onto the workbench, the bird perches nearby, one big dark eye cocked towards the willow rods, alert for anything that might move among them. When the bundle is untied, and falls open, the robin dives in time and again to capture insects, often coming within two feet of me as I work. Sometimes it perches on top of the bundle, and has learnt not to need to fly off when I pull out each handful of rods.

There is another singer somewhere up by the main gate, and every so often my companion is obliged to challenge it, flying across to the storeshed and darting out between the wooden slats at the back. I can't see what takes place, but it is not long before the robin is back in attendance. The harsh ticking-off I had received at first has now become a soft *tseep*, which seems to express a kind of contentment at our beneficial proximity.

It is the penultimate day of the month, and almost directly overhead in the woodshed a small drama is playing. A newly-hatched late brood of swallows occupies one of the nestboxes I put up five years ago. I had no idea it was in use until yesterday, when I found a half eggshell (white, blotched with dark brown spots) on the ground underneath the box. The young have hatched into an inhospitable time: gale-force winds and heavy rain are lashing Garronfield, which is why I am at work in the woodshed.

It seems almost miraculous that a swallow could find food in such conditions. Those insect-magnets, the cattle and sheep in the field, are nowhere to be seen. Presumably they are sheltering over in the lee of Barn Hill. The wind hurtles in gusts up the yard, roaring in the trees, and bouts of rain slap on the shed's metal roof. Yet time and time again there is a sudden flurry, an instantaneous peeping from the chicks, and another beakful of insects arrives, to be pushed into those gaping throats.

I can see three gapes, but sometimes the parent reaches past them to presumably feed another chick at the back, so there are at least four youngsters in the box. In the early afternoon there is a respite in the weather, but it is all too brief, and the wind picks up again, the rain becoming more persistent and even heavier than before.

It begins to dawn on me that I have not seen both parents at the nest together, or even in close sequence, which must mean there is only one adult to feed the chicks. That would be a hard enough task on the best of insect-full days, but in these conditions it seems impossible. I imagine a scenario where, in the last days of the female sitting on the eggs, the male succumbs to disease, or a predator. The female sits tight, the chicks hatch, and now she is faced with the imperative of keeping them alive.

Among the woodshed's roofbeams an existential crisis is unfolding. The female's forays into the storm take longer and longer. Sometimes she brings food, sometimes she comes back with an empty beak, and perches on the lip of the nestbox for a while, alongside her hungry brood. Then after a short rest she flips down and out into the battering wind and rain.

Towards the end of my working day I realise that she has been out for more than half an hour. Just as I am putting my tools away, she flutters in, without food, and sits by her youngsters once again. No end to the storm is forecast until tomorrow. I come to the sad conclusion that the chicks are unlikely to survive till morning. I go indoors tonight with a heavy heart.

Thirty-seven: Bellwether.

In early November a momentous event occurred here at Garronfield: a nuthatch came to feed on the peanuts. I saw it mid-morning on that sunny day, and my heart skipped. At US election time there is talk of 'bellwether states', meaning those whose popular vote can often be taken as a microcosm of the whole national result. Originally a bellwether was the leader of a flock of sheep, and a bell would be placed around its neck, so that the shepherd would know the flock's whereabouts. By transference the term came to mean the setter of a pattern or trend.

The nuthatch is the bellwether of habitat change here. The species is a woodland-dweller, feeding on insects and grubs under the bark of trees, and nesting in holes and crevices in mature trees. They are not far away from us: nuthatches are commonly seen in the woods on either side of the road at the top end of our track, and this is not the first time I have seen one at Garronfield. In the autumn of 2015 one briefly visited our feeders outside the caravan, and there was a single visitor in autumn 2016. It seemed likely that these visitations were young birds seeking a new territory, or the locals exploring. The trees we had planted a few years previously had scarcely grown to the extent they would be considered suitable habitat for nuthatches. And there were no more sightings, until now.

The nuthatch is one of my favourite birds. They are dapper, with slate-blue crown and back and wings, and russet underparts. They are somehow piratical in appearance as well, with a bold black eye-stripe and dark dagger beak. Their behaviour lives up to this appearance, for there is no sharing with other birds when a nuthatch is on the peanut feeder. The dapper pirate will guard its hoard against all-comers.

The species is one of a handful that have extended their population range in Britain in recent years. In the mid-1990s their furthest limit was the northern Lake District, and they were not found in Scotland. Yet slowly but surely they moved further north, and then westwards into Galloway. The reasons for this expansion are not known, but by 2011 there were occasional sightings in the Central Belt.

Two weeks went by since that first visit, without a further sighting. That is not to say the nuthatch didn't come to feed in that time, but if so, I didn't see it. I was beginning to think it was a one-off, and my claims about suitable nuthatch habitat misguided, or at best premature. But on a bright Sunday morning, as I did my regular bird-count for the BTO's Garden Birdwatch scheme, the nuthatch was back.

It made several visits over the course of half an hour, flitting between Spring Wood coppice and the feeders. This time it did not confine itself to the peanuts, but tried out the seed feeder, and the fat block - I make them from lard, hard fat left over from cooking, and seed, poured into plastic butter tubs and yoghurt pots. They hang from a hook, a little lower down than the feeders.

In that half-hour a hierarchy of assertiveness became apparent. On the peanuts, the nuthatch chased off all rivals, but on the seed feeder would tolerate a bird at the opposite portal. To my surprise, it would give way to a house sparrow, and on the fat block actually share the food with a male sparrow, until the sparrow decided to have the block to himself.

So, in ascending order: the coal tit is most timid, but because it does not seem to eat at the feeders, it manages to dart in wherever there is a space, grab a beakful, and zip off into the trees to hoard each morsel. Both blue tit and great tit will evict the coal tit if they find it in their way, and the great tit will do the same to the blue tit, and to the chaffinch as well. When the nuthatch turns up it drives the three tit species and the chaffinch away, but gives in to the house sparrow.

On the ground, a robin repeatedly drives a dunnock away. They are on adjacent pages in the bird-book, but are not of the same family. The robin is related to the blackbird; the dunnock is one of a small family of accentors. It used to be called, colloquially, 'hedge sparrow', but the correct name is dunnock. Confusingly, ornithologists now say the proper term is 'hedge accentor', but I persist with the shorter and more evocative name.

The dunnock is another of my favourite birds. The book says they have sparrow-like markings, but that is a lazy description. Study one closely and it is clear that the bold brown streaking on the back and belly is neatly arrayed, as though it had been combed into place. This is offset by a lead-grey crown, neck and chest. It is, like the nuthatch, a bird I would describe as dapper.

It annoys me when birdwatchers refer to 'LBJs' - little brown jobs. They mean the extensive range of species with superficially similar plumage, but it seems a rather dismissive attitude. Humans may struggle, with our poor eyesight, to distinguish one small brownish bird from another, but the birds themselves have no such problem. The sparrow and the dunnock, the skylark and the meadow pipit, may have similarities to confuse us, but they are species as distinct from one another as the kingfisher is from the jackdaw.

In fact, the LBJers may be somewhere on the spectrum of the ludicrous when it comes to ornithology. At the zenith of absurdity are the 'twitchers' who will travel hundreds of miles at the shortest notice to view a rare vagrant bird. I recall hearing of one such off-course migrant that had ended up somewhere in the Northern Isles. Word got about, and the place was soon besieged by men (mostly, it has to be said), with outsize optical equipment, all trained on the hapless bird which had taken refuge on top of a tall fence. The twitchers had a grandstand view as a sparrowhawk killed it and carried it off to eat.

The return of the nuthatch has restored my confidence. The trees we planted nine years ago have grown into young woods, so that Garronfield now consists of a mosaic of habitats: pasture, woodland, watercourses, wild scrubby areas, long grass, and outbuildings. The tally of bird species has grown to sixty-seven, mammal species twelve, amphibians four, butterflies nine. There are numerous moth species, but as yet we have only identified a handful, and there are damselflies. An incomplete count of plants exceeds fifty species, though we have stopped counting for the time being, until we have the leisure to look closer. The overall impression is one of abundance.

The diversity of species has increased dramatically over the past ten years. Many of the birds we see now would not have been interested in the place back then, especially the woodland-dwellers. Willow warbler and chiffchaff, blackbird and song thrush - these breed here now. Although reed buntings are still occasionally seen in the winter, they have gone elsewhere to nest, and the mallard whose nest I found among the newly-planted trees of Spring Wood has not returned. But these inevitable losses are offset by the gains.

The nuthatch will have been able to cross Garronfield from tree to tree, without having to fly more than a few yards across open ground. The red squirrel that came to visit us a couple of years ago will likewise have been able to cross the place within almost continuous cover, as our shelterbelts provide wooded corridors to move through. For the nuthatch, there aren't really yet any suitable nest-sites, except perhaps in the small stand of sycamores that were here when we arrived, but there is an increasing opportunity to feed. For the squirrel also, no place yet to build a drey, but the hazels are now producing nuts.

I would like to see a treecreeper here; I would like the tawny owls to come out of Bennel Wood and hoot from our trees; I would like the squirrel to become a regular visitor, even though it will, literally, cost us peanuts. But for now I am content to watch the bellwether nuthatch, the dapper pirate, flitting between the feeders and Spring Wood coppice.

Thirty-eight: April.

For T S Eliot, April was 'the cruellest month, breeding
>> Lilacs out of the dead land, mixing
>> Memory and desire, stirring
>> Dull roots with spring rain.
>>> (The Waste Land)

For us, this year - 2020 - April is the strangest month. A plague stalks the globe, and all is locked down. New rules apply: how to wash your hands, how often to go out, who you can and can't visit, and how far to stay from everyone else. Plans for a family get-together at Easter have been cancelled. It amounts to a coronavirus curfew.

In the days before Easter we have a run of fine dry weather, and the rather neglected allotment gets first-class treatment, readied for a new season of productivity. We walk around the woods, now unnervingly open since almost all the ash trees have been coppiced, their brash spread around the stools, a tangle of pale grey bones. Garronfield holds its breath ahead of the coming long exhalation of green life.

Yet the birds of this place do not hold their breath: everywhere the males are proclaiming their territory, and there is an excited pairing-up all around us. In March Wood a chiffchaff repeats his eponymous call; his cousin the willow warbler sends down his sweet fluting song from the top of a willow; a song thrush rehearses brief phrases, one of which sounds like a phone ringing.

One mid-morning a skein of high-flying grey geese heads due north, over Maiden Pap and Lotus Hill and on to Svalbard. I wonder what communication had passed between them in the hours since dawn, and what the trigger that had caused them to commit to their long migration.

Along the edge of March Wood I have been planting sixty long willow cuttings, to form pollards. They are a new variety here (*salix viminalis x Schwenerii* 'Tora', no less) and they have come from Geoff Forrest, an artist based up by Loch Doon. Every year he gets commissioned to make living willow sculptures, and 'Tora' is his favourite variety for this purpose, but in recent times he has struggled to buy in supplies, and his own growing is limited because of red deer in the forest around his place.

I first came to know about Geoff when a number of people, enquiring about my willow stock, mentioned that he had given them my contact details. Eventually I got around to emailing him a thankyou, and sporadic correspondence followed. He regularly participates in Spring Fling, the local open studios event over May bank holiday, and last year I noticed that he was going to share a studio by Loch Ken, much nearer for us to visit, and we resolved to make the time to do so.

Every once in a rare moon I meet someone who I know immediately to be a kindred spirit. We ducked out of the rain into the cramped lobby that Geoff had been lent to show off his wares, and while we waited to speak to him I took in the breadth of his work: fine willow bowls stained with blue shades, from eggshell to ultramarine at the rim; steel sculptures of hares' heads, so close to the real in detail you would almost think they might blink at you; a large basket filled with quirky creations that expanded the notion of basketry into more abstract forms.

At first there was a small knot of people in the studio, but I knew instinctively which was Geoff, and when the chance came I said: "I'm guessing you're Geoff - I'm Rob Drake." A warm greeting and a firm handshake were a good start, and straight away we were chatting, almost excitedly, about some of the things we held in common.

The rain eased off, a surge of visitors broke into the conversation, and Karen and I stepped outside, where Geoff was displaying his steel sculptures of running and leaping hares. They are elemental things, sketchlike in simplicity, yet containing all the vitality of the living creature. It suddenly struck me that something of this sort would be perfect for the gable of our plant room.

The plant room gable is the first thing you see when you approach the cottage. It doesn't have a door or window, because just behind the wall are the boiler and accumulator tank that provide our heating and hot water. To relieve the blankness of the stone facing I incorporated a feature in the apex: a ring of small slates with one large blue slate behind them - a sort of round blind window. And when I roofed the plant room, I fixed an

L-shaped metal bracket under the last ridge tile, with the thought that at some point a thing of beauty might live there.

And now I was looking at such a thing, and could imagine a hare leaping above the ridge, or better still, a pony - a variation on 'the cow jumped over the moon'. We went back inside, and broached the idea with Geoff. His fascination with hares had led to their being the main theme of his current work, but he had made sculptures of other creatures, and a leaping pony was not out of the question. It was something pleasant to ponder for the future.

Then our talk turned to willow, and the growing of it. As he became aware of the extent of our land at Garronfield, it was then that he asked if I'd be interested in growing some for him. It didn't need much thinking about: we had the space, and it could prove to be a worthwhile enterprise, of benefit to both of us. We didn't exactly shake hands on a deal, but it was a firm proposal for Spring 2020.

While we'd been talking, Karen had rummaged through the big basket, and had come up with an unusual piece of willow work, rather like a papoose, or a backpack, which she bought. There was a horizontal brace through the top rim, where you'd attach straps if you were going to wear it on your back, and this wasn't a piece of willow. Geoff asked if we knew what wood it was. It was rather driftwood-like, de-barked, pale in colour, with a series of indentations along the length of it, and I did know what it was, because I'd seen plenty of it at Garronfield.

Along the line of the burn, gorse grows in abundance. In the winter we cut some and throw it over the fence into Paradise Meadow for the ponies. They eat the spines, and particularly like the flowers when they are out. But then they strip the bark, and leave bare sticks similar to the one I was gazing at. "I think I do know - it's gorse". I got the impression that Geoff was simultaneously impressed and disappointed: I knew my stuff, but I'd deprived him of the chance to surprise us. Nevertheless, if there was any tangible thing to demonstrate a bond between us, it was that stick of gorse.

I meant to send Geoff a photo of the plant room gable, but never got around to it, and it wasn't until the new year that we started to talk again about willow growing. I'd paced out a suitable patch at the edge of March Wood, and there was room for up to sixty cuttings, spaced about half a metre apart. He was very busy, and February bled into March. I wondered if it was going to happen at all, but then came the news that he'd got the cuttings, and could deliver them on Friday afternoon. This was just days before the lockdown started, and Geoff proposed leaving the willow at our gate, without us having face-to-face contact.

I wasn't having any of it. I said that if he was OK to come here, I was happy for him to do so. At that stage the conflicting advice was for over-seventies to stay indoors, while the Ordnance Survey (an arm of government) was saying, no, get outside and keep yourselves fit and healthy - just stay two metres away from anyone else. I felt that Geoff and I, though neither of us quite in the age-bracket, might be able to manage that.

So in the end he rolled up in his van that Friday afternoon before the world changed utterly, and we walked about the place, chatting agreeably about many things. As Geoff admitted later, it was far better than leaving the willow at the gate and running away.

There are about forty stirks in Alan Carter's fields: they have been here since mid-month. As a single linear herd they rotate clockwise around Barn Hill: in the late afternoon they trudge northwards just beyond our fence. I looked to see if the same one was always in the lead, and it seems this is so: a black one with two distinguishing white patches on its left cheek. This one doesn't appear to be the tallest or heaviest of them, and as with the geese heading for Svalbard, I wonder what passes between these animals, that sets them off on their migrations.

But the cattle do not always rotate clockwise, it turns out. There is anti-clockwise movement also, which does not seem to involve the whole herd, and is not led by the black one with two spots - in fact it is more likely a brown one. I speculate that this might be a splinter group, literal mavericks, renegades refusing to conform. Somewhere out of sight behind the hill, however, they are evidently brought back into line, and come past, trudging northwards, led by black two-spots once more.

The stirks range from all-black to rich brown, with all manner of piebalds in between. They are scarcely pedigree, but seem content with their lot: day after day of warm sun and plenty of grass.

The month sets records for sunshine, lack of rainfall, and warmth. It adds to the tally of exceptional Aprils that have occurred since the start of the millennium. On the last Sunday evening, warm and still, a blackbird sings continuously from the elder below the cottage. About nine o'clock I open the kitchen door to hear him better, and go back into the living-room to turn off the CD of Debussy's *Suite bergamasque*, as the poor guy is, for now, hopelessly outclassed - even *Claire de lune* could not compete. The blackbird continues for another quarter of an hour, then seems to have stopped. I listen to the silence for a while. Then, on the edge of darkness, a few last sweet notes.

Thirty-nine: Making Hay While...

The sun shines in Galloway as well as it does anywhere in these temperate islands. In fact, some would say it does better here, especially along the coast, which in the past has brought artists to places such as Kirkcudbright, drawn by the quality of light - though this was more than just the direct rays of the sun, as beguiling at dawn and dusk and under cloud as on a blue sky day.

But where will there be a span of sunny rainless days to make hay? In 2020, after lockdown in late March there follows the sunniest April on record, then a May in which it does not rain at all until mid-month, and then only showers overnight. In these two months there were spells of fine weather that would have been perfect for haymaking - but of course there was no grass to cut. It was to be the 6th of August when I would go up on to Orchard Bank to open up our hay meadow.

The way we do things here, it requires four days at least to make hay. Five are better, and six better still, though this is asking a great deal from Galloway's weather. Not just sun, but a breeze is important, and from the right direction, because the beech hedge we planted along the top and down one side of the orchard is now tall and thick enough to shelter the bank from easterly winds. So, four back-to-back sunny days as a minimum, with a westerly breeze, in late June at the earliest, and more likely in July when the grasses have ripened, is what I look out for, almost obsessively.

Every morning I study the weather forecast. The Met Office is first call, then the BBC, whose data now comes from a different source, and then The Weather Outlook, which provides a 16-day forecast as well as shorter ones. Sometimes even the Norwegian forecaster YR gets a glance, particularly if the others are mismatching.

It seems to be becoming a pattern of reliable fine weather in spring, and then more changeable conditions through early summer. As the ideal time to cut the hay approaches, tension grows. It is a gamble to start, and if the choice to do so is ill-informed, then it is most likely a disaster. We have lost one crop entirely due to rain arriving earlier than expected, and we managed to save another year's cut by heaping it into two piles and sheeting them with tarpaulins until the weather brightened up again - but this is only possible if the hay is almost dry, as wet grass would soon heat up and go bad in a heap.

The 'window of opportunity' opens; the decision is made; the gear is waiting, ready to begin. In the right circumstances I like to 'open up' Orchard Bank the evening before. This entails going round the edge anticlockwise, cutting towards the middle, then round again clockwise, cutting towards the edge. This leaves an open strip that the first cut of the harvest will fall into. Opening up seems to help dry the new edge of standing hay, making for easier work the next day.

I cut the hay using a Stihl brushcutter with a 40-tooth saw blade. It is noisy work, and halfway through the day becomes quite tedious. Later on, though, as the uncut area shrinks perceptibly with every circuit, the spirits rise, and the last swathe feels like a triumph. Then I go away and try to forget about haymaking for a whole day.

The cut grass lies in the sun and shrivels: a brisk breeze will help to wick moisture out of the crop. By evening the bulk of the grass will have reduced considerably, and the top layer faded towards yellow. But underneath the hay will be green as when it was cut.

On the third day - mid-morning, or even later if there has been a dew - I take a wide spring-tine rake and begin to turn the hay. I work it uphill, into rows across the bank, with a narrow strip between rows. The rake turns the grass, but also lifts it into a loose heap, so that the wind can blow through, if there is any wind. It takes about an hour to turn the crop, and then it is left for two hours, until the rows are turned again, this time downhill. Later, a third turning rolls the rows uphill once more.

Each time the rakeful of grass feels lighter, but it is not yet hay. Nor is it hay the morning of the fourth day, but as the turning continues, and especially if there is both sun and breeze, the last vestiges of green vanish, and there comes a lightness and crispness that mean the harvest is ready to get in. I lift a handful of grass to my face and breathe in the scent that says 'I am hay'.

Up to this point the haymaking has gone along at its own pace. We can help by turning the grass as often as daylight allows, yet it is time and sun and wind that do most of the work. But now there is a bankful of hay ready to get in, and an urgency grips us. Out come the tote bags, or dumpy bags as they are also known, that previously have held the various sands and aggregates used in the building work. The hay is packed into them as tightly as possible, yet even when full they can still be manoeuvred. We work across the bank, leaving the bags standing, like giant multicoloured building blocks. When all the hay is bagged up we drag them to the bottom of the bank, and cram four at a time into the skip of our ancient dumper. I drive up to the barn, Karen climbs into the hayloft, and between us we heave the bags into the loft, which can take twenty-five of them, and if there's more than that they are stacked on the barn floor, to be used first.

As the last bags come in, a wave of relief and gratitude floods through me. I park the dumper, put tools away, and we go inside for an evening of celebration. There is a Gaelic phrase, *latha math* (laa maa), which means 'a good day' - or a momentous one, or auspicious. At some point before bedtime we will raise small glasses of malt - Highland Park, or perhaps Glenlivet - and say those words to one another. It has indeed been a good day.

Forty: Nutting.

On a calm dull Saturday afternoon in mid-September we go hazelnutting. Last year, on the advice of the Woodland Trust's website, we gathered them earlier while they were still green, and ripened them on a tray in the kitchen cupboard. This was only partially successful, and many kernels turned out to hold nothing more than the shrivelled remnant of a nut. So this year we have waited until they have ripened on the hazels, and we seem to have timed it just right, as a few clusters have already released their nuts, and those we pick come away easily.

You have to 'get your eye in': at first it is all too easy to fail to spot the clusters among the foliage, and even when you think you have adapted you still miss some. Back in the summer the bright green clusters were easy to see amongst darker green leaves, but now the nuts in their frilly husks are mottled brown, as is the tree's foliage, so they are camouflaged. It is almost as though they are aware that in their ripe state they will fall prey to predators, and invisibility is their best chance of survival.

But they do not succeed. The strap of my collecting-bag bites ever deeper into my shoulder as an hour passes, and needs emptying when we have gone through March Wood and part of Spring Wood, before we return to cull the rest. In our original planting we put hazel around the fringes of the woods, and in small groups by Holly Grove and Corner Stones. This has made harvesting easy, and in less than two hours we have gathered the crop.

Of course we have missed some, and these will be bounty for squirrels and badgers and voles and field mice. It is probable that only a small proportion of nuts will be left to sink into the soil, to germinate and produce new growth. But over the coming winter I intend to coppice many of the hazels, cutting out the original stem, which will stimulate the growth of side-shoots. As well as hazelnuts, I hope in time to have a harvest of straight hazel sticks.

Probably more than any other tree species, hazel has shaped human civilisation since prehistoric times. It was an early coloniser, after birch and willow, as the last ice-sheet retreated northwards. The abundance of hazel pollen in the record indicates that the species was not overshadowed by rivals, as when that occurs hazel fails to thrive, and stops flowering. There must have been extensive hazel woodlands after the ice, and it is known that Mesolithic people ate quantities of hazelnuts.

Early in Neolithic times humans discovered the usefulness of natural coppice. Where large and unworkable trees had blown over - or been pushed over by large beasts - the regrowth was often slender and straight, and could be cut down and worked with stone tools. This is the origin of woodmanship, which has continued for at least 6000 years.

The clear evidence of early woodmanship lies in the wetlands of the Somerset Levels, where trackways were constructed out of coppiced hazel, as well as other species, to allow access across soft ground. Later trackways were made out of woven wattle hurdles, using huge numbers of hazel rods grown within a sophisticated coppicing system designed to produce rods of exactly the same thickness, meaning that the coppice would be of different ages. It is probable that the rods were a by-product of growing hazel to provide leaf-fodder for livestock. Hurdle-ways continued to be made throughout the Bronze and Iron ages.

Hazel was usually the predominant species employed in medieval wattle-and-daub, the method of construction that used woven wattle panels to infill a timber frame, which would then be coated with clay or dung to provide a substantial wall. Hazelnuts were a crop worth having in the Middle Ages, and the gathering of them was a labour duty for farming tenants, who would give the crop to their landlord to discharge part of their obligation.

By the early nineteenth century a tradition of nutting was well-established, and was an important social occasion in the rural calendar. However, the sociability side of the occasion gave rise to complaints from some landowners about townspeople descending upon hazel-woods and using the pretext of harvesting nuts as an excuse to have a party in the woods in the evening, complete with drinking and carousing. But surely the harvesting of any crop ought to be cause for celebration.

In more recent times hazel has been in decline, and nutting as a social event no more than a memory. The decline may partly be due to a lack of woodmanship, but a contributory factor is predation by grey squirrels and possibly woodpigeons. The squirrels strip the trees of unripe nuts - which is why the Woodland Trust advised beating them to it - and any not eaten are not ripe enough to grow, leading to a loss of regeneration of hazel.

Of all the produce we generate here, hazelnuts have the longest shelf-life. It was only a matter of days before this latest harvest that I cracked the last few kernels of the previous one. The ability to store food for a time when little else is available has been key to human survival through milleniums. When mankind does eventually manage to attain its last and most terrible achievement, total annihilation, I wonder if there might come a final year when the remnant of our species survives for a short while with hazelnuts an item on an ever-dwindling menu.

Forty-one: The Thinning of the Leaves.

In late October a steady leaf-fall in the coppice allows 'beyond' to appear once more. The outlines of the ponies move behind a waving, thinning screen; the steep field rising to New Farm is visible, and the farm buildings along the ridge; the forested skyline of Bainloch Hill can be seen once again. The interior of the coppice becomes exposed, and the activity of birds more obvious. It is as though the curtains are being drawn aside in a theatre where the actors are rehearsing their winter play behind the scenes.

There is quite a *stooshie* (that useful Scots term meaning a disturbance, a fuss, a to-do): chaffinches, dunnocks, and especially robins, chasing one another through the trees, flaring up, holding ground, striving for dominance, staking claims. One of this year's robins broke its neck against one of our kitchen windows, no doubt being chased by a rival. There are certainly two left, and possibly three, but as robins do not care to keep company with one another it is hard to tell. At dusk they tick one another off, sounding like the winding of an antique clock. Daily they are to be seen around the house and among the outbuildings. I say *chip-chip* when one is nearby - that is the way to talk to robins.

Leaf-fall also exposes a winter's work. At the end of November I will cease to be a builder, until next spring, and will instead become a woodman. I will put away the grey and black work clothes and the blue overalls, and fetch out the woodcutter's motley of greens and browns. My work boots will no longer be stained with mortar and stonedust, and will become clagged with mud and woodchips.

There is the annual willow harvest to undertake, and I plan to start that earlier than usual, in December rather than January, and bring the harvested bundles into the woodshed, so that I may process them under cover on wet days. And then the heavier work of coppicing will begin.

Already I am looking forward to getting started, and as I walk about the place I am daily more able to see into the dense regrowth, and find myself making mental note of which areas will be cut this time around. Last winter the work we carried out was unusual, because ash dieback had become apparent throughout the woods the previous summer. We decided then to coppice almost all the ash trees, in both woods, in the hope that vigorous regrowth might help to stave off the virus. The consequence of this plan was that some stands of other species did not get cut, and they are now ripe for coppicing. So it will be a small patch here, another there, in both woods.

As well as this primary coppicing, I will revisit areas cut three and four years ago, in order to cull the willow regrowth. We interplanted with willow cuttings during the early years of establishing the woods, and the vigorous growth helped to provide shelter for the other trees. But this winter it will be time to start thinning the willow, so that alder and aspen, oak and hazel may have more space to spread their branches.

Another intended harvest will be of the gorse that grows naturally along Caulkerbush Burn. This is a most useful scrub tree: it makes excellent firewood. The branches grow as a series of straight sections divided by bends, like elbows. Thinner branches snap easily at these joints, when dry, and a supply of kindling is soon produced without the use of any tool.

But otherwise tools are an essential part of coppicing. If I were asked, what are the three most important inventions of the last hundred years, the list would never quite be the same all the time, because things evolve, but the chainsaw would always be on it. Anyone whose life has been free of the requirement to cut and process timber could not possibly comprehend how this one tool has done away with so much backbreaking labour, and speeded-up the work immeasurably. I have used chainsaws for the past forty years, and although in other realms I am happy with the traditional ways of hand-working, I could not easily imagine a return to the axe and the double-handed saw for felling trees.

In the early days I had an assortment of second-hand machines of different makes, until I could afford a Husqvarna saw, and stuck with that make for several years. But eventually when Huskies became too expensive I converted to Stihl, and have been faithful ever since. They are powerful, reliable, and durable chainsaws: I am only on my second one in eighteen years, and it is still working well. For round about twelve years, from 2001 onwards, the sixteen-inch bar on my saw would nose its way through the tough fibres of large trees - oak, beech, elm, ash - converting them to firewood, and ripping down the sawn rounds whenever these were too hard to split with an axe. The machine was always powerful enough to deal with the knottiest of timber, and work through trunks twice the diameter of the saw's bar length.

It is some years since I felled a large tree. The last ones were the dead elms on the Glen estate, that Robert the farm manager would let me take for firewood, when we lived at Paradise Cottage. My chainsaw work now is exclusively among the small-diameter timber of our coppices, and I am thinking of buying a smaller and lighter saw, with a shorter bar than the one I've got. It is time to talk to Bryan, the forest equipment man in Dalbeattie.

In the end it is very simple. In these days of buying online, and never being entirely sure that you're getting the right thing, it is refreshing to talk to someone who knows the gear they are selling, and can reassure you that the equipment will suit your purpose. A short phonecall, and it's all sorted out: another Stihl saw, with a twelve-inch bar, and all for less than £200. "Aye, it'll do you just fine", says Bryan. I tell him I'll pick it up tomorrow. "Aye, nae bother".

Powerful and versatile as the chainsaw might be, it can't carry out all the tasks involved in coppicing. After the poles have been felled and cut into roughly six-foot lengths, and once the whine of the saw and the yellow fountains of woodchips have ceased, a peace returns to the coppice, and in it just the *snack-snack* of the billhook or the machete as it completes the work. There is something calming about the quieter task of snedding, neatly removing the smaller side branches from the poles, and spreading them across the ground between the stools.

But once the firewood lengths have been carted away and stacked, and the brash has been spread evenly across the ground, it is time to leave the coppice alone. There may well be no footfall at all in the cut for years to come. At first the brash, to deter deer, deters us also. Then the regrowth quickly becomes impenetrable. In summer the interior of the coppice is unseen, masked by the foliage of trees along the margins. Only in late autumn can we begin to peer into the middle, at the thinning of the leaves.

Forty-two: A Visitation.

It seems we have been adopted by one of the most beguiling creatures among British wildlife. Almost every day, usually in the late afternoon or early evening, a hare appears, close to hand. This is not the steel work of art mentioned in a previous chapter, but the real living thing: the brown hare, *Lepus europaeus*. It often sits at the bottom of Orchard Bank, just above the side door, and remains there, quite unperturbed, as we come and go along the path to the door.

It seems unconcerned to follow the normal habits of the species. Whilst it may be active at night for all we know, it is not purely nocturnal, being in open view in broad daylight. It is not wary of us: one afternoon in late May I spotted it nibbling some of Karen's seedlings in pots outside the polytunnel, and went out to dissuade it. It made off, casually, doing a stop-start circuit of the polytunnel, which brought it back round towards me. It came to within three feet, before hopping away along the edge of the track.

It does not seem to make a form - the shallow depression hares scrat out to hide up in during the day. It is noticeable, however, that this individual has a preference for lying-up in patches of dead, brown vegetation, where it is perfectly camouflaged. In fact, part-way through writing the first paragraph of this chapter I glanced through my study window, and there was the hare, sitting in full view, grooming itself in the middle of a patch of dry grass. I would have been spooked by this, if I were susceptible.

The first time the hare came close was in late April. For the past three years Karen has cultivated a patch of ground outside the sitting-room window to grow flowers, a colourful oasis in the midst of an otherwise building site. This time she was part way through digging the patch over, had heeled the fork in at the edge of her digging, and come inside to make the tea. I was in the midst of getting a fire going, and stepped over to look out of the window as the flames took hold. It took a moment to register, but there, little more than two yards away, crouched next to the fork on the brown mat of last year's flowers, was a hare.

It was the eye turned towards me that gave it away: the grey-brown of its fur blended with the vegetation so well that it was almost invisible.

But that wide round eye, with its dark centre, seemed to look right into mine. I had never been quite so close to a wild hare, and I was thrilled. I beckoned to Karen, and we watched as it nibbled a stalk or two, then dozed off for a while. We spoke in whispers, and moved slowly, but it was clear the hare could neither see nor hear us. I took photographs, the camera close up to the glass. In some of the images the hare almost fills the frame, every detail exquisitely clear.

This was not the first time we had seen a hare at Garronfield. When we first had the place there was an occasional sighting, then none. Perhaps the place was too busy and noisy during the main phase of construction. But then we began again to see a hare, infrequently, in the fields, and a few times I chanced across forms in the lengthening grass. It was always a single creature, and seemed to edge closer to us as time went by. We would see it at the top of Orchard Bank on occasion. Then in mid-June last year two leverets appeared, in Karen's flower-garden, lurking among the blooms along the edge of the track, about five or six yards away from the house. After that there were more regular sightings of single individuals.

To this point I have used the impersonal pronoun, but I cannot stop myself from thinking of the hare as *she*. This is similar to my instinctively referring to a fox as *he* - for this I can hold Mr Tod and Reynard responsible, but there is no corresponding influence for the hare. It is completely illogical: as parthenogenesis does not occur in wild mammals, our hare is as likely to be a buck, a Jack-Hare, as a doe. There is little physical difference between the sexes, though Warnes' 1921 *Pocket Guide to Animal Life of the British Isles* asserts that 'the Jack-Hare has a smaller body, shorter head and redder shoulders than the Doe.' This is contradicted by more recent studies, which claim the male to be up to five per cent larger than the female.

Apart from the leverets, I have not seen two hares together here, so comparison is impossible. The solitary individual could be a buck, without a female nearby to mate with, but this seems unlikely. If the creature is a doe, then she ought to be giving birth to a succession of litters, from February through to September, meaning there is at least one male in the vicinity, and ultimately a number of offspring round and about. This raises a question: how can I be sure that the hare I see is always the same one?

I can't. Until I see two hares together, or two individuals in different places at the same time, uncertainty continues. I am intrigued by the possibilities, however. One is that the hare is a doe, and she is keeping away from the secret night-time form, near which nestle two or three leverets in their own small forms that she has scraped for them. At dusk she will go to her form, and the youngsters will gather there to suckle for a short while, before they disperse once more, the doe leaving by means of a tremendous leap, in order not to leave a scent trail back to the young.

Another possibility is that this is a buck, having performed his coital duties, completely uninvolved in the rearing of young, loafing about the place, nibbling on tasty leaves and flowers, waiting for the leverets to be weaned and to make off on their own and for the doe to be ready to mate yet again. A tough life, as they say.

The third possibility is that I am seeing more than one hare, maybe one of them a well-grown leveret from early on in this mild spring, or perhaps both buck and doe, who circle the place in their own orbits, keeping out of the way of one another - as Warnes' Pocket Guide says: 'the Hare is as solitary and retired as a hermit'. The end of May is approaching: perhaps the summer months will uncover the mysteries.

My immunity to being spooked may be getting thin. The day after I began writing this chapter, and the hare appeared when I hadn't even completed the first paragraph, I was working on it again, and had reached a stopping-point, which might have proved to be the end of the chapter. I saved what I'd written, put the file away, and closed my laptop.

I sat looking out of the window, trying to think if there was anything I'd left out of my brief account of the life of hares. Within moments the hare appeared, loping along the track that runs in front of the cottage. I had grown used to seeing the creature from my study, since it seemed to have abandoned Orchard Bank in favour of the area of rough grass between the track and Top Field's fence-line, so I wasn't startled when it appeared. But instead of continuing up the track, it suddenly veered towards the cottage, heading straight for my window. I stood up to see it better, and it was sat still, just a few feet away. The hare's nose was pointing in my direction. I had the slightly uncomfortable feeling it was staring at me.

The rational mind tried to have its say. I was sure I couldn't be seen through the glass, and I felt that if I *could* be seen, the hare would turn its head so that one eye could scrutinise me better. But...but... Why on earth had it come to my window at just that moment? There was nothing on the bare ground for it to eat: it seemed as though the hare had deliberately turned aside, that it had come *to speak to me*. And then I thought of what I had left out of my account, perhaps deliberately: the place the hare occupies in folklore and superstition.

Folklore ranges from the eminently practical (e.g. When is the best time to cut a hazel wand to make a stick? Answer: when you see it...), through to acts and beliefs of ignorance and cruelty, such as Hunting the Wren, mentioned previously, and the old practice of killing a hare to place its body in the start of a new building, presumably to confer good luck (but not for the hare). We are supposed to be living in a more enlightened age, but a certain strand of ignorance and cruelty continues, threaded into the dark underbelly of modern life.

According to Stefan Buczacki's *Fauna Britannica*, the brown hare has been for long ages the subject of superstitious beliefs. Fishermen and miners would turn back from going to sea or to the pit if they saw a hare on their way to work. It was believed that if a pregnant woman saw a hare, her child would be born with a 'hare lip', the facial deformity of a split in the upper lip. And hares, as with cats, were associated with witches, to the extent that a witch might take the form of a hare to escape pursuers, or so it was thought.

It seems that the amount of myth and superstition surrounding hares is out of proportion to their place in the natural world. They are not commonplace: most people will not see a single hare from one year to the next. For the main part, they are secretive, hidden in the long grass of everyday life. When they do come to our attention, they seem to be doing odd and unaccountable things, such as a pair 'boxing' in spring, or several hares racing in line across a field. We know now that these behaviours are part of the mating ritual, with a doe fighting off the attentions of a buck, or a squad of males chasing a female who is ready to mate. In times before these facts were understood, hare behaviour was considered 'mad' and otherworldly.

Underlying superstitious beliefs about hares is the mistaken fear that the creatures have bad intent, that they are able to confer ill fortune on us, that to even see one presages miscarriage, deformity, disaster, death. The demons of imagination - goblins, trolls, imps, brownies - are all ugly beings: it is odd that one of our most beautiful and graceful creatures should have been associated with malevolence. Yet as I stared at the hare through my study window, and it seemed to stare back, I felt a visceral quiver, a sense of trepidation at the animal's strange but opportune behaviour, and recognised that if I had lived in an unscientific age, I might have been afraid of what the hare had come to say to me.

If the hare had indeed come to speak to me, then what might the message be? If the creature were an ambassador for the wild life of Garronfield, she would speak of two things: fertility and fragility.

At the time of writing this, in early June, she may well be suckling her second litter of young, or they might even be making their own way in the wild world, and she is looking to mate for a third time. We won't see the whole of this output - perhaps only one or two of the young will reach maturity - but it is another demonstration of nature's optimism, its outpouring of life upon life to confer survival on each species.

And the abundance is hard to miss just now. A flurry of starlings, and here are the parents, scuttling about under the feeders to poke morsels into the open beaks of at least five fledglings, who are just learning to fend for themselves, but are still willing to be fed if the easy food is available. Another flurry, this time of house sparrows, and it is impossible to know how many young there are, because as soon as one is fed by a parent it hurtles away out of view, while a couple more come in to land, and be fed in turn. This same day I see the cock sparrow mate with the fluttering hen, below their nest-site in the storeshed, where the next clutch of eggs will be laid.

Yet for all the optimism, the abundance, the seeming robustness of wild nature, the hare speaks also of fragility. Normally timid, living in open habitat, her only defence against predation a turn of speed and the ability to jink away at right angles, her young especially vulnerable to silage-making machines.

At present the brown hare is listed as 'least conservation concern', although population numbers have been declining in Europe for half a century, so the level of concern may change at some point. But what about the decline in house sparrows, greenfinches, song thrushes, the loss of pollinators, the almost extinction of red squirrels? An unthinking person may feel they have no need for greenfinches, but that would be a foolish mistake.

The hare outside my window demands that I *think*: that I make myself aware, as never before, of the fine balance that on the one hand allows nature to thrive, and on the other leads to decline. And she insists that I *see*: that in the mosaic of habitats at Garronfield there is biodiversity, such as there has not been here for many generations, and that I am keeper of it.

She tells me there is a place for her here, in the summer grass we leave long for winter grazing. She tells me there is a place here for many others: for sparrows and starlings, frogs and lizards, roe deer and voles, wasps and black beetles, orchids and knapweed. We may have done some things to help, but much has just *happened*, of its own accord. The message of the hare makes my heart glad: I am not afraid of what she has come to say.

It is hard to remember what the place was like on that first day when we visited, and my blood began to sing. I have to dig out the photographs to recall the empty field, the rushes, the derelict cottages beneath the bank. Now, on a morning walk, I take the meandering path through Spring Wood, emerging at the bottom end to stop by the gate into Paradise Meadow. I look back up the way I have come, and I can't see my house, because a wood stands in the way. At moments like this, when I realise what we have, my blood sings again.

Forty-three: Of Lightness and Weight.

It was August 1993, at Alnmouth on the Northumbrian coast. The day is carved in my memory, not for the place itself, lovely though it might be, but because of one seemingly trivial detail that had an affect on me.

Karen and I had driven down to the shore, and walked along the beach and back. As we drove out of the carpark we passed a mobile home set up on a grassy place. Sitting in the sun, in their folding chairs, were an older couple, a table between them, and under it a bristle-haired terrier. I don't recall them reading, or talking to one another. I have the impression they simply sipped their tea and watched that day's happenings as they passed by.

I've said the scene is carved in my memory, but I'm not sure if I said anything to Karen, then or later. What abides is the potent thought that went through me, that this would be a way to live - to not have ties, to be unmoored from responsibilities, to be free to follow the weather south, or north, or east, or west.

My marriage had failed the year before, and I had run away from the daily responsibility of home and family. I was twelve months into my new relationship with Karen, and, though very much in love, I don't think either of us were yet absolutely sure we were destined to be together always. I had a sense of freedom, such as I had not known since I finished college and started work. I could feel possibilities bubbling away somewhere.

The freedom was an illusion, of course. I had not completely abandoned my family, and continued to support them, as well as trying to have as much time with my children as was possible. In fact, I was probably more tied at that moment than I had ever been, but illusions have a way of worming themselves into the rational and making the untenable seem real.

Some dreams bubbled to the surface and could be spoken of. One was the idea to sleep on every inhabited island around the coast of Britain and Ireland. To travel by motorhome would assist in some, but not all, of such an itinerary. In the following years we did get to sleep on islands - Inishmore, Lindisfarne, Islay - without the aid of a motorhome. Gradually the dream withered away to a husk: Karen and I were settling into a life that rooted us, and there were fewer chances to 'just head off'. What chances occurred were never spontaneous: in the absence of that mobile living space on wheels we always had to book ahead for ferries and accommodation.

Five years on from that remembered day in Northumbria we acquired an eight-month-old rescue dog, a border collie we named Sky, after a dog called the same we'd encountered on Islay. Once he had settled himself with us, we went away for a weekend on two occasions, leaving him in a good kennels nearby. The first time he seemed excited to be in a new place with other dogs around, probably not associating it with the other kennels we had liberated him from a year previously. He was ecstatic when we turned up to collect him.

But the second time, he knew. He remembered the place as we drove up the track, and his head went down. The girl in charge was kind, and took Sky out for a walk while we sneaked off. Two days later, going to pick him up, I chanced to see him in his pen before he saw me, and I knew from his look of abject misery that we were never going to leave him there again.

So a further bond, albeit a pleasant one, had arrived. I had stood at an illusory crossroads in 1993, and looked down the ways to left and right, that seemed to offer some kind of freedom. But then, as it turned out, I carried straight on along a road of responsibility, and burden, and weight.

In his 1984 novel *The Unbearable Lightness of Being*, Milan Kundera wrote of the fifth century Greek philosopher Parmenides, who saw the world divided into pairs of opposites: light/darkness, fineness/coarseness, warmth/cold, being/non-being. He called one half of each pair positive (light, fineness, warmth, being), the other negative. But when it comes to the pairing of lightness and weight, which is positive, which negative? Parmenides' answer was that lightness is positive, weight negative. Kundera questioned whether this was right, and pointed to the underlying ambiguity.

I have never been satisfied with this positive/negative paradigm. I have little mind to be a deep philosopher myself, but I do know that silk may be very fine to wear in the evening, around the house, yet it is utterly unsuited to the harsh wear and tear of manual labour. In similar vein I get pleasure, in roughly equal amounts, from ice-cubes in my lemonade and from a bowl of piping-hot lentil soup. I think I will leave being/non-being to one side for the moment.

Darkness does not have a negative aspect in my view of the world. I can't recall ever being truly afraid of the dark. I have one childhood memory of feeling terrified at some night-time monster that hovered in a corner of my bedroom, but it was just a damp patch on the wall, and when I was shown what it was, the terror vanished. It wasn't the darkness that frightened me: then, as always, it was my own imagination.

But then we come to lightness and weight, and, as Kundera suggests, here lies ambiguity. A heavy heart is an empty heart, a heart that is light is full of love and joy. That motorhome on the Northumbrian coast was a vision of lightness, a dream of islands. Instead, I have come to rest on the heavy soil of Galloway.

Kundera felt that the heaviest of burdens crush us to the ground, so that as we become closer to the earth, our lives become more real and truthful. On the other hand, an absence of weight allows us to soar free and become only half real. He asks which we should choose, weight or lightness?

If it seems as though I have chosen, I would have to say I am not crushed. There is the weight of my physical intimacy with every square inch of Garronfield, yet my heart is full and often light, sometimes almost unbearably so.

If my being consists of both weight and lightness, at the same time, but not always in equal measure, then I must have an image, a symbol on which to hang the delicate clothes of this interior life. If I have searched a long time, and ventured far in my imagination, the answer has always been near. There is one entity in which weight and lightness exist in balance: it is a tree.

Forty-four: Roots.

'Roots' is the title of a lino print by the sculptor and artist Elizabeth Waugh. I bought the print from her at her Bakehouse Studio in Langholm in May 2007, and it has hung on four walls since then. At Paradise Cottage it hung on the chimney-breast above the fireplace, and each evening whenever I lit the fire I would study the image as the flames were taking hold.

The background is a landscape: it is unlikely to be a particular place, but rather an archetype, yet it definitely has the feel of being Scottish. At eye level, as it were, there is a small lake, or lochan. This is surrounded by high ground: rounded hills clothed with conifers to the right; sharp crags to the left. At the base of the crags, by the water's edge, is a small clump of trees, which although delineated by minimal lines and marks, are indisputably willows.

At the foot of the lochan there is low-lying undulating ground, which half-conceals a small cottage with a round window in the facing gable and a lean-to against the other. There is a suggestion of plough-marks near the building and perhaps the remnants of a hedge.

In the very foreground lie the tumbled stones of a wall, which are almost disappearing out of the bottom of the picture. It is as though this boundary had been pushed aside, toppled by the sheer strength and vitality of the main feature of the linocut, which the eye is drawn to at first sight, and then returns to again and again.

A tree. Right at the heart of the image is a tree, its trunk dead centre, with the start of the larger limbs and a few smaller branches visible at the very top. Clearly it is autumn, as the branches that can be seen are almost bare, and a few heart-shaped leaves are caught midway in their journey towards the ground. This tree is quite old: the trunk is gnarled and fissured. The lower end of the trunk divides into strong ribs with dark crevices between them, and these ribs form the juncture with the most extraordinary element of the scene.

Trees usually have their roots tucked away out of sight below ground, but this one has every root on display, a network of fibrous fingers across a rocky tummock, whose crown is gripped with palpable force. These roots are not delving into the earth, but spreading across it, winning the battle with obdurate stone, and threatening to obliterate the cottage and its land.

Where trees grow on rock, their roots are inevitably close to the surface or even above it, and the image recalls instances where an individual tree seems to exist, even thrive, without apparently any soil to grow in. To see a tree growing in such adverse conditions brings to mind the human notion of survival, which is to create conditions for life in situations that would normally be inimical to existence. Even in favourable circumstances clearings must be made in the forest, boulders moved aside, wood chopped for the fire, and water carried. Humans have always done this, and although most of us are far removed from having to clear the land to make fields, we still carry within us the urge to make a place for ourselves on the earth.

The lino-print tree, growing on rocky ground, is a symbol of that fierce tenacity which has led humans to dig deep and to reach high, in a way that no other creature on earth has done. The tree is an image of the vitality and optimism of life itself, a powerful sinew between sky and earth. And inevitably it is a symbol of sexual potency. Probing roots break open rock and weather grinds it to a paste of soil, in which a seed can find nourishment enough to put out a green shoot.

There is one final aspect of the lino-print to mention, and that is the light. In the vee of the high ground there is a strong brightness. It could be dawn or sunset, but somehow I have a preference for it being sunset. The sun itself is not visible, perhaps hidden behind the trunk, but more probably this is the kind of sky where cloud is thick enough to conceal the sun's outline, yet thin enough to make sunlight diffuse, bright enough to hurt the eye. And the nature of this light is to get around corners, as it does in this case, silvering the left side of the trunk.

There is that kind of light as I write this, if I glance out of my study window towards the crest of Bennel Wood. Elizabeth Waugh's lino-print hangs in its final resting place, in the cottage that is home for the remainder of my life. The view from my window is of the upper part of Top Field, and beyond, the nearest trees of March Wood, that year by year have been putting down their roots into the peaty and stony ground set aside for them. Just as oak and alder, aspen and holly take hold of the land, I too am putting down roots, in a way I never have before. The trees I have planted are doing it for me: fixing me in place, making me take sustenance from this bit of ground, telling me to grow as straight and true as I am able.

And it is with the image of each tree, and every tree, that the ambiguity of weight and lightness is resolved. A tree presses itself to the earth, clutching the ground with tenacious fibres, all the while sending up branches, thinner and thinner, that will in season bear a crop of leaves, to flutter in the wind all spring and summer, then fade to dry skeletons, and fall, feather-light, to nourish the hungry roots of the tree that gave them life.

I look from the lino-print's white sky to the brightness beyond the ridge, and back again. In the picture a darkness gathers above the shining. There may be a storm on its way; it is autumn; the tree is old. But its roots grip the earth and will not loose hold.

Also

A SOLITARY TRADE - the Art and Craft of Dry Stone Walling.
Bookcase, Carlisle.
ISBN: 9781904147381.

BECKS AND GILLS of the Northern Fells.
Bookcase, Carlisle.
ISBN: 9781904147596.